Electrolytes

A Systems Guide to Hydration, Electrolyte Balance, and Physiological Stability

Nathan Reed

PERENNIAL PUBLISHING

Published by Perennial Publishing

Paperback ISBN: 978-1-971996-24-0
Hardcover ISBN: 978-1-971996-25-7
E-book ISBN: 978-1-971996-26-4

Contents

Introduction

Electrolytes are often introduced as simple companions to hydration. They are described as substances added to water, lost through sweat, or consumed to support performance, recovery, or general well-being. This framing is widely accepted because it is easy to observe and easy to apply. Fluids are consumed, fluids are lost, and electrolytes appear to follow that same pattern. Within this view, their meaning is defined by intake and replacement. They are treated as materials that enter the system, are used or depleted, and must be restored to maintain function. This interpretation is not entirely incorrect, but it is structurally incomplete. It reduces a foundational regulatory system to a surface-level exchange.

The limitation of this intake-based model becomes clearer when it is examined against how physiological systems actually behave. Intake describes what enters the system, but it does not explain how that material is distributed, how it is retained or excreted, how it participates in signaling, or how it interacts with other variables across compartments and time. A given quantity of sodium or potassium does not carry a fixed meaning independent of context. Its significance depends on gradients, on ratios, on distribution between intracellular and extracellular environments, and on the regulatory state in which it is being used. When electrolytes are viewed primarily as consumable inputs, these relationships are obscured. The system is reduced to a container rather than understood as a coordinated structure.

This reduction leads to predictable interpretive errors. Stability becomes associated with sufficient intake, and disruption becomes associated with deficiency or excess in a narrow sense. The system is assumed to behave linearly, where more input leads to more stability and less input leads to less stability. Variation in symptoms or system behavior is then interpreted through this same lens, as though each outcome could be traced directly to a single variable crossing a simple threshold. This approach offers clarity at the surface, but it does not account for the complexity of physiological coordination. It fails to explain why similar intake patterns can produce different outcomes under different conditions, why symptoms often emerge in patterns rather than isolated events, and why stability can be maintained for a time even when underlying conditions are no longer aligned.

To move beyond this limitation, electrolytes must be reframed. They are not best understood as substances whose primary significance lies in their presence or absence. They are better understood as elements of signaling and regulation. Their function is expressed through gradients—differences in concentration across compartments—that create directionality, electrical potential, and osmotic structure. These gradients are not static. They are maintained actively, shaped continuously by transport mechanisms, fluid movement, hormonal regulation, and system demand. Electrolytes therefore operate less like stored materials and more like components of an active infrastructure through which physiological systems communicate, coordinate, and maintain stability.

This infrastructure is distributed rather than localized. Sodium, potassium, and magnesium participate in different aspects of system behavior, but their significance emerges through interaction rather than isolation. Sodium shapes extracellular conditions and fluid partitioning. Potassium anchors intracellular stability and membrane behavior. Magnesium influences modulation and control of signaling and activation. These roles are not independent. They are interdependent, forming a network in which each element influences the conditions under which the others function. Stability arises from the coherence of this network rather than from the adequacy of any single component viewed in isolation.

Within this framework, imbalance must also be redefined. It is not simply the absence of a substance or the presence of too much of it. It is a disruption of coordination. Gradients may narrow or become distorted. Distribution may shift in ways that alter pressure or compartmental integrity. Signaling may become less predictable or less stable. Ratios between electrolytes may drift, changing how systems interpret and respond to internal conditions. These disruptions may occur even when total quantities appear sufficient, because what matters is not only what exists, but how it is organized and regulated within the system. Imbalance is therefore relational rather than purely quantitative.

This relational view also explains why symptoms do not map cleanly onto single causes. Fatigue, dizziness, irregularity, or muscular dysfunction do not necessarily indicate a simple deficiency. They are expressions of system behavior under altered conditions. They reflect how gradients, distribution, signaling, and regulation are interacting at a given moment. Their meaning emerges through pattern and context rather than through isolated interpretation. A symptom is not a direct instruction. It is a signal that coordination has changed. Understanding that change requires attention to the system as a whole rather than to individual variables alone.

The perspective developed in this book follows this systems-based approach throughout. Each chapter examines a specific aspect of electrolyte behavior, but no aspect stands alone. Gradients are connected to distribution. Distribution is connected to signaling. Signaling is connected to regulation. Regulation is connected to adaptation and context. The aim is not to produce a list of facts about individual electrolytes, but to clarify how these elements interact to create or disrupt stability across physiological systems. The structure is therefore cumulative. Each concept builds on the previous one, not by adding isolated detail, but by expanding the network of relationships that define system behavior.

This approach also defines what the book does not attempt to provide. It does not offer intake recommendations, supplementation strategies, or performance-oriented guidance. It does not frame electrolytes as tools for achieving specific outcomes or optimizing particular functions. These forms of interpretation rely on the same input-based model that the book seeks to move beyond. Instead, the focus remains on explanation. The goal is to make the underlying mechanisms visible so that electrolyte behavior can be understood in terms of coordination, regulation, and stability rather than in terms of isolated intervention.

The expectation of direct solutions is therefore intentionally set aside. Systems do not become stable through isolated actions applied without regard to context. They become stable when their internal relationships remain aligned under the conditions they are exposed to. This alignment cannot be reduced to a single rule or prescription because it depends on variables that change across time, environment, and system state. The value of a systems-based understanding lies not in providing immediate answers, but in clarifying how the system behaves so that its responses can be interpreted more accurately.

This shift from input to infrastructure changes how electrolyte physiology is perceived at every level. It moves the focus from what enters the system to how the system organizes itself. It replaces linear cause-and-effect assumptions with relational and context-dependent reasoning. It reframes stability as an emergent property of coordinated interaction rather than as the result of isolated sufficiency. And it positions electrolytes not at the periphery of physiological discussion, but at its structural core, where gradients, signaling, distribution, and regulation converge to make coherent function possible.

Understanding electrolytes in this way requires a change in perspective rather than an accumulation of facts. It requires attention to relationships, to patterns, and to the conditions under which systems maintain or lose coordination. This introduction establishes that shift. What follows develops it in detail, examining the mechanisms through which electrolyte-driven infrastructure supports stability and the ways in which disruption emerges when that infrastructure becomes strained.

Signal vs Substance

Electrolytes function as signal carriers, not consumable inputs. That distinction is not semantic decoration. It is the structural difference between treating physiology as inventory and treating it as coordination. When electrolytes are understood primarily as substances to be possessed, replaced, or accumulated, their role appears materially simple: a set of minerals present in tissue and fluid, available in measurable quantities, entering and leaving the system through familiar channels. That view is not entirely false, but it is incomplete at the level where physiological order is actually maintained. Electrolytes matter not because they merely exist inside the organism as stored material, but because their charge, placement, concentration, and movement permit systems to communicate across space and time. Their significance lies in the fact that they are electrically active participants in regulation. They establish differences, and those differences make signaling possible.

The operative unit is not the ion in isolation. It is the gradient. A charged particle suspended in fluid becomes physiologically meaningful when it is unevenly distributed across compartments, membranes, and tissues in a way that allows force, movement, and information transfer to occur. Sodium is not important simply because sodium is present. Potassium is not important simply because potassium exists in a measurable amount inside cells. Magnesium is not important simply because it can be listed among essential substances. Each becomes functionally relevant when its charge contributes to a relationship of separation, tension, and regulated movement. Physiology depends on controlled asymmetry. The system does not seek equalization everywhere. It depends on non-equality being maintained in specific places, at specific magnitudes, within specific ranges, for specific purposes. Electrolytes are among the principal means by which that non-equality is constructed and defended.

It moves the discussion away from consumption and toward organization. It also changes the meaning of imbalance. If electrolytes are imagined primarily as substances, then imbalance is easily interpreted as insufficiency, depletion, or missing material. The correction seems conceptually simple: restore what is absent. But if electrolytes are understood as signaling elements, imbalance becomes more complex. It may reflect displacement rather than disappearance, compartment distortion rather than total loss, altered movement rather than inadequate supply, or failure of regulatory coordination rather than failure of intake. A system can contain the relevant ions and still behave as though communication has degraded, because functional integrity depends on where charge is held, how gradients are maintained, and whether signaling architecture remains coherent under load.

This is why electrolyte behavior cannot be reduced to a hydration story. Fluid and electrolytes are inseparable in many contexts, but they are not identical subjects. Hydration language tends to flatten the distinction by treating ions as accessories to water balance, useful insofar as they help fluid enter, remain, or circulate. That framing captures one visible surface effect while obscuring the deeper logic. Water follows osmotic forces. Osmotic forces depend on dissolved particles. Those particles alter compartmental behavior. But the same ions that influence fluid distribution also govern excitability, membrane behavior, contractility, conduction, transport, and intracellular coordination. To describe electrolytes only through hydration is to describe electrical architecture through one of its downstream manifestations. It is a real manifestation, but it is not the whole mechanism.

At the cellular level, this becomes especially clear. Cells do not function as chemically uniform sacs. They maintain boundaries, and those boundaries are electrically meaningful. The membrane is not simply a barrier separating inside from outside. It is an interface across which gradients are established, held, and selectively altered. Sodium and potassium exist on opposite sides of this interface in different concentrations, and that arrangement is not incidental. It is a condition of readiness. The membrane potential, excitability, and responsiveness of the cell emerge from this unequal distribution. When channels open, the movement of ions across the membrane is not merely a change in local chemistry. It is a change in signal state. It alters what the cell can do, how it responds, and how it coordinates with neighboring structures. Electrical signaling is not layered on top of electrolyte behavior. It is one of the principal expressions of it.

That point is often obscured by the ordinary language of nutrients. Nutrients are commonly understood as materials used for structure, energy, or synthesis. They may be stored, metabolized, transformed, or incorporated into tissue. Electrolytes overlap with that world only partially. They are certainly substances, and some are obtained through intake in familiar ways, but their central role is not well described by the logic of fuel or building material. Their primary physiological significance lies less in what they become than in what they permit. They enable gradients, and gradients enable regulation. Their role is therefore relational and dynamic rather than simply compositional. They are less like raw material waiting to be used and more like distributed conditions required for communication to remain possible.

This is why the term presence can mislead. But function depends on far more than existence. A sodium ion in extracellular fluid is not interchangeable with a sodium ion in the wrong compartment at the wrong concentration relative to surrounding structures. A potassium ion inside a cell contributes to one form of order; the same ion in excess outside the cell can alter excitability and destabilize signaling. Magnesium may be physically present within tissue while its functional availability, binding state, or system role shifts under stress in ways that change coordination without requiring dramatic total depletion. The material has not vanished, but the signal environment has changed.

This relationship between material presence and signal function is one of the most commonly misread features of electrolyte behavior. Surface interpretations tend to collapse everything into quantity because quantity is easy to imagine and easy to measure. But living systems are not governed by the most intuitive layer of description. They are governed by controlled disequilibrium. Charge separation, concentration differences, and selective permeability create the possibility of order. What matters is not simply whether an ion exists in the system, but whether its distribution still supports the signaling architecture required for coordinated function.

The language of consumption tends to obscure this architecture because it frames electrolytes as if they were discrete inputs with direct, linear effects. Yet many disturbances do not follow that line. A shift in fluid compartments can change concentration without substantially changing total content. Hormonal signaling can alter retention, excretion, and movement across membranes in ways that change electrical behavior without obvious changes in intake. Stress can alter intracellular and extracellular distribution. Acid-base shifts can alter ionic relationships. Changes in autonomic state can influence transport and responsiveness. Under these conditions, the problem is not best described as absence. It is better described as altered regulation of signal conditions.

This is one reason electrolyte language often becomes conceptually unstable in public interpretation. The category appears familiar enough to encourage oversimplification, but its actual role is too distributed to fit neatly inside a single explanatory frame. It touches fluid balance, yet it is not reducible to fluid. It touches nutrition, yet it is not well described as ordinary nourishment. It appears measurable, yet its function often depends on relational states not captured by totals alone. It appears concrete, yet it operates through invisible gradients and electrical contingencies. As a result, interpretation frequently defaults to the most tangible version of the subject: substances in, substances out, balance as replacement. That framing is easier to narrate than signaling integrity, but it distorts the object under examination.

To understand why, it is useful to separate inventory from function. Inventory refers to what is materially present in the system. Function refers to what that presence is actually doing within a coordinated network. A system may possess substantial material inventory and still lose functional integrity if compartment relationships deteriorate. Conversely, a system may temporarily preserve signaling function under changing inventory conditions through regulatory compensation. Endocrine signaling, renal handling, membrane transport, and intracellular buffering all participate in this compensation. The visible surface can therefore remain deceptively stable while the regulatory burden required to preserve that stability increases underneath. When electrolyte behavior is treated only as material accounting, this compensatory layer disappears from view.

That disappearance matters because it changes how disturbance is interpreted. If the model is consumptive, disturbance looks like deficiency. If the model is signaling-based, disturbance looks like altered coordination. The first asks what is missing. The second asks what relationship has changed. A deficiency model narrows attention toward replacement logic. A coordination model widens attention toward compartment behavior, transport control, membrane conditions, endocrine regulation, fluid distribution, and electrical responsiveness. One sees a lack of substance. The other sees a disturbance in organized difference. In many physiological states, the latter is the more accurate description.

This becomes especially important when the system is under load. Stress, fasting, heat exposure, sleep disruption, prolonged exertion, and illness all alter fluid handling, hormonal signaling, and cellular behavior.

Under these conditions, electrolyte behavior often shifts not because the system has suddenly become empty of key ions, but because regulatory priorities change. Distribution may be defended differently. Excretion may change. Retention patterns may shift. Intracellular and extracellular relationships may be reweighted. The system may preserve function in one domain by narrowing margin in another. This is not random fluctuation. It is adaptive prioritization under constraint. To understand what electrolytes are doing in these states, one must observe how signaling conditions are being preserved, distorted, or redistributed, not merely how much material appears to be entering or leaving.

This also clarifies why the term balance can be misleading when used too casually. But electrolyte stability is not stasis. It is active regulation around narrow and meaningful ranges, often through continual adjustment. A stable signal environment does not imply stillness. It implies controlled oscillation without loss of functional integrity. Ions are constantly moving. Channels open and close. Pumps consume energy to preserve asymmetry. Hormonal systems adjust renal handling. Fluid compartments shift in response to osmotic conditions. Cellular states fluctuate. Stability therefore emerges not from immobility, but from managed motion. Electrolyte order is dynamic and energy-dependent.

That energy dependence is itself revealing. Maintaining gradients requires work. A cell does not passively remain excitable. It expends energy to preserve the ionic differences that make future signaling possible. This means electrolyte function is bound to broader regulatory capacity. Signal integrity depends not only on the ions themselves, but on the machinery required to place them correctly, keep them separated, and move them selectively. Membrane pumps, transporters, endocrine responses, and organ-level handling all participate in this effort. If any of these layers become constrained, signaling quality may degrade even if the relevant substances remain physically present. Again, the problem is not simply inventory. It is the system's ability to preserve meaningful electrical order.

This is one reason simplistic input-output thinking fails so quickly once the subject is examined with sufficient resolution. Electrolytes are often described as though they enter the body, circulate as helpful substances, and then support hydration or function in a relatively direct way. But in reality, what matters is the fate of distribution. Once present, ions are sorted, exchanged, retained, excreted, buffered, shifted, and compartmentalized according to an ongoing hierarchy of regulatory priorities. It continuously arranges them. That arrangement is the physiological event. Without it, the ions are materially present but functionally incoherent.

The phrase signal versus substance therefore identifies a difference in explanatory center. It denies that substance is the correct primary lens. If the chapter were framed around material alone, the resulting model would remain too blunt to explain excitability, transport, conduction, or the instability that emerges when gradients collapse despite continued physical presence of the underlying ions. The signal frame is more accurate because it places emphasis where physiology itself places emphasis: on charge relationships, selective movement, and distributed coordination.

This shift in lens also helps explain why electrolytes should not be interpreted as ordinary consumables. Consumables are usually understood through acquisition. They are obtained, added, and used. Their meaning begins at entry. Electrolytes do enter the system through intake, but their meaning does not begin or end there. Once inside, their role is defined by placement and interaction. They are not physiologically valuable simply because they have been ingested or absorbed. Their value is realized only when they are integrated into gradients that preserve signaling coherence. That coherence is maintained through active regulation, not passive possession.

To say that electrolytes are not nutrients in the traditional sense is therefore not to remove them from physiology's material world. It is to resist placing them inside the wrong explanatory category.

Traditional nutrient language tends to emphasize amount, sufficiency, storage, and supply. Electrolyte physiology requires a different vocabulary: distribution, excitability, membrane behavior, osmotic force, electrical asymmetry, transport, and signaling precision. These are not merely advanced details added after the basics. They are the basics. If the foundation is built around intake and quantity alone, the rest of the subject becomes distorted from the beginning.

This distortion becomes especially pronounced when intake is equated with stability. Stability is not a direct output of material addition. A stable system is one in which gradients remain appropriately organized, signaling remains coherent, fluid compartments remain functionally distributed, and regulatory systems can respond without excessive strain. Intake may contribute to that under some conditions, but contribution is not equivalence. Stability is an achieved state of coordinated regulation, not a receipt of consumed material. The same input can support stability in one context, fail to alter instability in another, or become functionally irrelevant if the disturbance lies in distribution, retention, or regulatory control rather than supply.

A problem appears, a substance is named, and the relationship is drawn in straight lines. But physiological systems rarely operate in straight lines. They operate through contingent relationships, threshold behavior, compensatory shifts, and cross-system interaction. Electrolytes participate in these interactions at such a fundamental level that they are often mistaken for simple variables precisely because they are everywhere. Their ubiquity makes them seem obvious. In reality, ubiquity is part of what makes them structurally important. They are involved not as isolated agents but as enabling conditions for system-wide coordination.

Sodium illustrates this well. It is often described as a hydration-associated mineral, but its deeper significance lies in its role as a dominant extracellular cation shaping osmotic distribution, volume relationships, and electrical conditions across membranes. Potassium is often discussed as a dietary mineral associated with general cellular health, yet its significance lies centrally in intracellular electrical architecture and excitability. Magnesium is often described in broad and vague terms as supportive or calming, but physiologically its role is more exacting: it participates in enzymatic function, stabilizes electrical processes, and modulates signaling environments in ways that exceed any simple substance narrative. The ion matters because of the relationships it sustains.

Once that is seen clearly, electrolyte imbalance can no longer be interpreted as a single category of absence. Imbalance is a breakdown in organized difference. It may arise through depletion, but it may also arise through redistribution, failed retention, inappropriate compartment shifts, altered hormonal handling, impaired transport, or overload within a signaling environment that cannot maintain coherence. The word imbalance is therefore useful only if it retains this structural complexity. If it is reduced to lack, it ceases to describe the system accurately.

This is also why electrolyte discussion cannot be confined to episodes of obvious dysfunction. Their role is not limited to moments of collapse. They are continuously involved in maintaining ordinary physiological order. Nerve conduction, muscle contraction, glandular signaling, cardiac rhythm, intracellular communication, and fluid compartment behavior all rely on conditions that electrolytes help establish. Their importance is therefore not episodic but architectural. They are part of the baseline arrangement through which coordination remains possible at all.

When a system component is constantly present and constantly active, it becomes easy to interpret it as background material rather than active structure. Electrolytes often disappear into this background because they are familiar, small, and chemically unremarkable to non-specialist language. But their physiological role is not small. They are among the principal means by which living tissue maintains electrical readiness, fluid partitioning, and coordinated responsiveness. They are less analogous to supplies and more analogous to the conditions under which organized function can continue.

To understand electrolytes accurately, then, requires abandoning the reflex to ask only how much is present. Quantity matters, but only inside a wider architecture of placement, relation, and controlled difference. That is where electrolyte physiology actually lives: not in isolated amounts, but in gradients; not in material possession, but in regulated asymmetry; not in hydration slogans or intake narratives, but in the continuously maintained conditions that allow systems to communicate without collapsing into noise.

Gradients

Concentration gradients drive cellular behavior and system function. Physiological movement does not begin with substance alone, and it does not emerge from material presence in the abstract. It emerges when concentration differs across space in a way that creates directionality. A gradient is not merely a numerical inequality between one location and another. It is an organized difference that permits force, movement, signaling, and regulation to occur. Without that difference, ions may still exist, fluid may still occupy compartments, and tissues may still retain material content, but the system loses one of the central conditions required for coordination. Function depends not on evenness, but on maintained asymmetry.

The importance of this point becomes clearer when the organism is understood not as a continuous undifferentiated fluid mass, but as a system of compartments separated by boundaries that matter. Intracellular and extracellular spaces are not just anatomical labels or descriptive conveniences. They are distinct regulatory environments with different concentrations, different electrical properties, and different functional roles. The interior of the cell is not meant to mirror the fluid surrounding it, and the fluid surrounding it is not meant to collapse into the interior. Physiological behavior depends on the fact that what is concentrated outside is not concentrated inside to the same degree, and what is concentrated inside is not distributed equally outside. This separation is not disorder. It is design at the level of regulation.

When sodium is predominantly extracellular and potassium predominantly intracellular, this is not a passive chemical accident or an incidental byproduct of tissue composition. It is a structured arrangement with consequences. Magnesium, while more diffusely discussed and more variably bound, also participates within this logic of compartmentalized regulation. These ions are not simply dissolved substances floating in fluid. Their placement contributes to the creation of different internal and external environments, and those different environments establish the conditions under which movement becomes meaningful. A cell surrounded by a medium identical to its own internal composition would not be more stable in any useful sense. It would be less capable of signaling, less capable of selective exchange, and less capable of maintaining electrical identity. Uniformity would not represent completion. It would represent collapse of functional distinction.

This is the point at which many interpretations begin to drift into error. Whole-system averages are often treated as if they describe physiological reality more accurately than compartmental structure. They do not. A total amount of sodium within the body does not explain what sodium is doing across extracellular space relative to intracellular space. A total amount of potassium does not explain excitability if the relationship between compartments has shifted. A measured quantity can obscure the fact that what matters is placement. The system does not operate according to total possession alone. It operates according to differential distribution across boundaries.

A gradient, then, is best understood as a stored directional condition. It creates the possibility that movement will occur in a particular way when permeability allows it. It is not motion itself, but it is the reason motion becomes structured rather than random. If one compartment contains a higher concentration of an ion than another, and a pathway opens between them, the resulting movement is not arbitrary. It follows the direction created by that difference. Movement in physiology is not generally the expression of substances wandering without logic. It is the expression of constraints and differences interacting through controlled access. The gradient is what gives movement its orientation.

That orientation is essential because physiological systems do not merely require movement. They require patterned movement. It is not enough for ions to move. They must move in ways that preserve timing, sequence, responsiveness, and electrical integrity. The opening of a membrane channel, the depolarization of an excitable cell, the transfer of charge across a tissue, the shift of fluid between compartments, the transport of solutes across boundaries—these are not meaningful simply because something changes position. They are meaningful because the change occurs along an established gradient, with directionality that is already encoded in the structure of distribution. The gradient contains the possibility of action before the action occurs.

This is why intracellular and extracellular environments must be understood as more than physical locations. They are differently composed fields of potential. The cell membrane does not merely separate them. It preserves their difference and selectively permits communication between them. That selective permeability is only physiologically useful because the compartments are unequal to begin with. If they were equal, opening a channel would have little consequence. No meaningful force would be released. No directional movement would be available. No electrical event would unfold in the way physiology requires. The membrane is therefore not important only as a barrier. It is important because it allows a structured difference to be held until movement is appropriate.

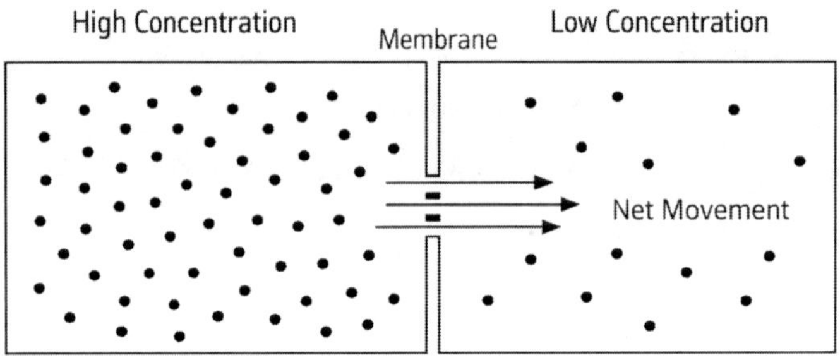

[Figure 1: Unequal ion distribution across compartments creates directional movement, establishing gradients as the source of force and signaling.]

Once this is established, the meaning of movement can be clarified with greater precision. Ion movement is often described in ways that make it sound spontaneous, as though particles simply travel through fluid according to generalized tendencies. But in physiological systems, movement is not meaningful because it occurs. It is meaningful because it occurs in relation to gradients, boundaries, and regulatory conditions. An ion does not move simply because movement is possible. It moves because concentration differences, electrical forces, permeability states, and membrane conditions together create a directional event.

This distinction matters because it separates physiological motion from conceptual noise. If ions are imagined as drifting through a generalized fluid environment, then distribution appears secondary and compartment boundaries appear incidental. But if ions are understood as participating in structured asymmetry, then movement becomes intelligible as a regulated process rather than a random one. Sodium moves inward under certain conditions because a gradient exists that makes inward movement energetically favorable when access is granted. Potassium moves outward under certain conditions because its distribution creates a different directional tendency. These tendencies are not arbitrary properties of the ions themselves. They are relational consequences of how the system has organized them.

Directionality is therefore not an optional refinement added after the fact. It is one of the central meanings of a gradient. It is a difference that creates a preferred path of movement. In this sense, concentration is not only descriptive. It is functional. Higher and lower concentrations do not merely tell us where more or less of something is located. They establish what will happen when the system allows interaction across that difference. Physiology depends on these latent directional relationships being preserved in advance of use.

This preservation is especially important because many regulatory events are fast. Electrical signaling in excitable tissue does not pause to construct its own driving forces from scratch. The force is already there in the gradient. The signal is possible because the difference has been maintained ahead of time. This is one of the reasons gradients are more important than quantities alone. Quantity can describe how much material exists. It cannot by itself explain what the material is capable of doing. A system may contain a large total amount of an ion, but if the distribution of that ion no longer supports a directional force across relevant boundaries, its functional meaning changes dramatically. Capacity depends on asymmetry, not just abundance.

A whole-system inventory can suggest sufficiency while obscuring functional failure. It can also suggest loss where the more relevant issue is redistribution. If sodium shifts in a way that alters extracellular concentration relative to intracellular space, the functional implications are not captured by asking how much sodium the organism contains in aggregate. If potassium leaves the intracellular space and accumulates where it changes membrane behavior, the issue is not merely whether potassium exists in total. The issue is that the gradient governing excitability has been altered. Averages flatten what physiology depends on.

This flattening is one of the most common interpretive failures in discussions of electrolytes and fluid regulation. It treats the organism as though it were a single pool rather than a structured arrangement of distinct but communicating compartments. Under that model, any description of content appears sufficient. But the body is not regulated as a single pool. It is regulated through selective separation. Each compartment matters not because it contains fluid, but because it contains fluid differently. Those differences establish force. Force establishes movement. Movement under control establishes function. Once the compartmental structure is ignored, the logic of the entire system becomes harder to interpret accurately.

This is also why equilibrium is not the correct physiological ideal. It suggests balance, calm, and correction. But in systems governed by gradients, complete equilibrium across compartments would not represent successful regulation. It would represent the disappearance of directional force. A membrane separating two environments of identical composition would not preserve signaling capacity simply because the system had become equalized. It would instead lose one of the main conditions required for excitability, transport, and coordination. Controlled asymmetry, not uniformity, is the relevant form of order.

The physiological asymmetry is not disorder. It is structured difference held within meaningful bounds. The system is not trying to eliminate all inequality. It is trying to preserve the right inequalities in the right places at the right magnitudes. This is why the language of "balance" becomes misleading when it implies sameness rather than coordination. Electrolyte stability is not the absence of gradients. It is the preservation of gradients that allow communication and controlled exchange to continue without collapsing into noise or stasis.

This distinction becomes especially important when considering what it means for a gradient to fail. Gradient collapse is not merely a drop in quantity. It is the loss of meaningful difference between compartments. When that difference narrows beyond functional range, the consequences extend beyond concentration alone. Electrical behavior changes. Membrane responsiveness shifts. Signaling becomes distorted or blunted. Directionality weakens. Exchange becomes less ordered. Fluid behavior may change as osmotic relationships shift. In short, function deteriorates not because material has necessarily disappeared, but because the system can no longer preserve the structured asymmetry required for coherent operation.

Depletion is one possible mechanism of disruption, but it is not the only one, and often not the most conceptually useful first description. A gradient can collapse because ions are lost, but it can also collapse because ions redistribute in ways that erase the functional difference between compartments. It can collapse because membrane control fails, because transport becomes constrained, because fluid shifts alter concentration relationships, or because regulatory systems no longer maintain separation with sufficient precision. In each case, the problem is not only that there is less of something. The problem is that there is less directional structure.

A system can contain substantial material content and still behave as though signaling capacity has narrowed. The total amount may not explain the observed dysfunction because what has changed is not inventory alone, but gradient quality. Gradient quality refers to whether the difference across compartments still exists at a magnitude sufficient to produce reliable movement and response. When that quality degrades, the system becomes less capable of translating stored asymmetry into controlled action.

The consequences of that degradation vary depending on where the gradient matters most in a given context, but the underlying logic remains consistent. Cells lose clarity in their signaling environment. Excitable tissues become more vulnerable to instability. Transport relationships become less efficient or less directional. Compartmental boundaries begin to lose some of their regulatory meaning. The system becomes harder to coordinate because one of the principal organizing conditions—maintained difference—has weakened. Collapse, in this sense, is not dramatic because it is visible. It is significant because it removes the directional architecture on which function depends.

This is why gradients must be understood as actively maintained structures rather than passive states. It is easy to imagine a concentration difference as something that simply exists until disturbed, like water resting at different heights in connected containers. That analogy can be useful at a very early stage, but it fails quickly if it implies that gradients are naturally stable without cost. In physiology, gradients are not merely inherited conditions. They are continuously defended arrangements. They require work. The system must actively preserve the fact that one compartment remains different from another, even though passive physical forces would tend to erode that difference over time.

This active maintenance is one of the most revealing aspects of electrolyte physiology because it shows that gradients are not accidental. The organism spends energy to keep them in place. Membrane pumps, transport systems, selective permeability, renal handling, endocrine signaling, and intracellular buffering all participate in preserving compartmental differences that would otherwise dissipate. It is central infrastructure. The system treats gradients as worth defending because so much else depends on them. Electrical responsiveness, fluid partitioning, cellular signaling, transport behavior, and tissue coordination all rely on asymmetries being maintained against the natural drift toward equalization.

From a purely physical standpoint, concentration differences tend to smooth out when pathways allow exchange. Left entirely to passive processes, compartments would move toward a less differentiated state. Yet physiology cannot afford universal equalization. It requires selective equalization in some moments and selective resistance in others. This means the system is not simply permitting physical laws to unfold unchecked. It is working within them to preserve useful difference. That is one of the defining features of regulated life: not escape from physical principles, but organized management of them.

Once this is understood, the meaning of maintenance becomes more precise. To maintain a gradient is not just to keep a number within range. It is to preserve directional capacity. A stable sodium gradient, for example, is not merely a stable concentration. It is a preserved condition of future inward movement under appropriate membrane conditions. A stable potassium gradient is not merely an intracellular abundance. It is a preserved condition of future outward movement under appropriate electrical and permeability states. The gradient is therefore a form of stored regulatory potential. It is not static because it is inert. It is static only in the sense that it is held in readiness.

This readiness helps explain why gradient disruption often precedes visible failure. Because the gradient is an enabling condition rather than a surface event, its deterioration may begin before overt dysfunction becomes obvious. A system can still appear operational while the margin of directional stability narrows underneath. Compensation may preserve surface behavior for a time. Membrane transport may work harder. Hormonal systems may alter handling. Fluid compartments may shift to protect one domain at the expense of another. But compensation does not eliminate the importance of the underlying gradient. It only demonstrates how much regulation is required to preserve it.

This is another reason whole-system averaging remains inadequate. It cannot show how much work the system is performing to preserve local asymmetry. It can report aggregate content while missing the regulatory burden required to maintain functional separation. A number can remain apparently stable while the architecture sustaining that stability becomes increasingly strained. That strain matters because gradients are not just descriptive features of the system. They are operational structures. If the system must expend progressively more regulatory effort to preserve them, then functional integrity is already under pressure even before overt collapse occurs.

Electrolytes are no longer understood primarily as substances that happen to move through fluid. They become participants in structured differences that generate direction, force, and response. The question is no longer simply whether ions are present, but whether their distribution across compartments still produces the asymmetry required for signaling and coordinated movement. This reframing immediately displaces several common errors. It weakens the assumption that quantity alone explains function. It removes the idea that equilibrium is inherently desirable. It resists the simplification of disruption into mere depletion. And it restores boundaries and compartments to their proper place as organizing features rather than incidental containers.

What emerges from this view is a more accurate picture of physiological order. Function does not arise because matter is present everywhere in sufficient amount. It arises because matter is distributed unevenly in ways that create meaningful directional relationships across selective boundaries. Those relationships are then preserved, modulated, and used by the system to coordinate behavior across scales. A gradient is therefore not a detail within physiology. It is one of the principal forms through which physiology becomes possible. Without gradients, there is no stored direction. Without stored direction, there is no controlled movement. Without controlled movement, signaling loses structure, and coordination loses force. The system remains materially populated, but it becomes functionally less distinct, less responsive, and less coherent.

Fluid Distribution

Electrolytes determine where water goes. Fluid distribution is not organized primarily by the volume of water entering the system, but by the osmotic structure created by charged particles distributed across compartments. Water does not impose order on electrolyte placement. It responds to the order electrolytes create. Wherever concentration differences emerge across a boundary that permits water movement, water shifts in relation to that difference. This is the central principle underlying fluid partitioning in physiology. The question is not simply how much fluid exists in total, but where that fluid is held, how it is distributed between compartments, and what electrolyte arrangements are directing that distribution at any given moment.

This immediately changes the meaning of hydration. Hydration is often interpreted as a matter of intake, as though fluid balance were determined mainly by how much water enters the organism over a defined period. Physiological fluid behavior does not operate through a single pool. It operates through multiple compartments separated by boundaries and regulated by concentration differences. Water is not simply present in the body in general. It is partitioned into spaces with different functional meanings, and that partitioning depends on electrolyte distribution. When hydration is reduced to water consumption, the spatial logic of fluid regulation disappears, and with it the central mechanism governing whether fluid is actually located where functional stability requires it to be.

The first correction, then, is to establish direction properly. Water follows electrolytes, not the reverse. This does not mean water is passive in every sense or that it has no regulatory importance of its own. It means that the directional force governing its movement between compartments is created principally by dissolved particles, especially electrolytes, whose concentration differs across relevant spaces. Osmotic behavior arises because water moves in relation to solute concentration. Where solute concentration is higher, water tends to move toward that space if the intervening boundary permits it. It is directional movement governed by asymmetry. Electrolytes create the asymmetry. Water responds to it.

That directionality is essential because fluid balance is not a matter of random exchange. It is a matter of controlled distribution across intracellular and extracellular compartments, each with different compositions, different pressures, and different physiological roles. The intracellular space is not simply fluid inside cells. It is a regulated environment in which concentration relationships support cellular volume, signaling, and functional integrity. The extracellular space is not merely everything outside cells. It is a distinct domain shaping interstitial conditions, vascular volume, transport relationships, and tissue-level exchange. Water moving between these compartments is therefore not just fluid relocating in an abstract sense. It is a shift in the distribution of volume across distinct regulatory environments.

This is where the concept of osmotic balance becomes central. Osmotic balance does not mean static equality everywhere. It means regulated fluid distribution across compartments whose electrolyte concentrations differ in specific and meaningful ways. Water movement occurs because compartments are not identical. If intracellular and extracellular environments contained the same concentration of osmotically active particles in exactly the same arrangement, there would be no directional basis for water movement between them. But physiology depends on non-uniformity. Compartments maintain different ionic profiles, and those profiles create osmotic relationships that influence where water is retained, where it exits, and how volume is stabilized. Osmotic balance is therefore not the elimination of difference. It is the managed interaction of differences.

Equalization across all compartments might sound like resolution, but complete undifferentiated equalization would erase the directional structure that makes compartmental regulation possible. Osmotic stability depends on controlled asymmetry, not uniform flattening. Water must remain appropriately distributed across different spaces, and those spaces must remain compositionally distinct for that distribution to retain functional meaning. A cell is not stable because it contains water. It is stable because water content, solute concentration, and membrane-regulated exchange remain coordinated within a defined compartmental structure. The same is true for extracellular space. Stability depends on preserved relationships, not on generalized sameness.

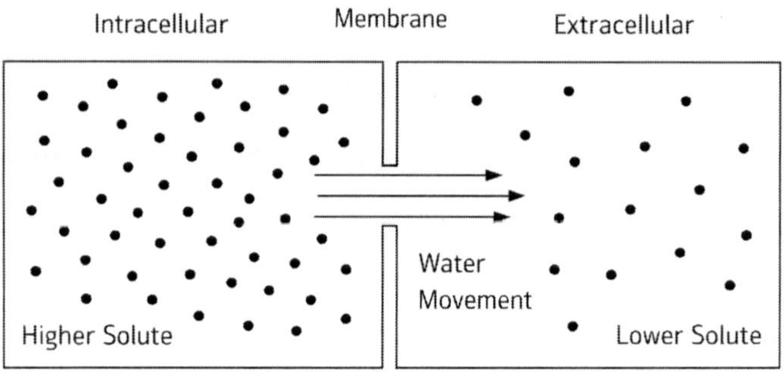

[Figure 2: Fluid movement across intracellular and extracellular compartments driven by osmotic gradients created by unequal electrolyte distribution.]

Once osmotic balance is established as the governing structure, fluid movement can be understood with greater precision. Water does not distribute itself according to abstract need or generic hydration status. It moves in response to gradients. If extracellular sodium concentration rises relative to intracellular conditions, osmotic forces shift in a way that alters water distribution across the membrane. If intracellular solute concentration changes, water responds accordingly. The movement is not random, and it is not primarily a direct expression of water intake. It is a response to the arrangement of osmotically active particles across compartments. This is why fluid regulation cannot be understood by volume alone. Volume without location is physiologically incomplete.

That incompleteness is one of the reasons fluid imbalance is so often misinterpreted. A person may appear fluid-loaded in one compartment while another compartment is relatively constrained. A tissue may accumulate fluid in a way that reflects distributional distortion rather than restoration of balance. Water may be present in substantial quantity and yet fail to support functional integrity if it is held in the wrong place relative to osmotic conditions. When fluid status is reduced to total intake or total presence, these distinctions disappear. But the system does not regulate a total in the abstract. It regulates partitioning. Partitioning is what determines whether extracellular volume is supported appropriately, whether cells maintain structural coherence, and whether pressure relationships remain within ranges compatible with coordination.

Sodium occupies a central position within this extracellular logic because it is the principal determinant of extracellular osmotic structure. Its predominance in the extracellular compartment is not incidental. It is one of the main reasons that compartment holds the volume it does. Sodium shapes the osmotic environment outside cells, and water distribution follows that environment. When extracellular sodium concentration changes meaningfully, extracellular volume behavior changes with it. It means that sodium distribution is one of the dominant variables influencing how much water is retained within the extracellular space and how strongly that space exerts osmotic influence relative to the intracellular environment.

This extracellular role extends beyond simple fluid presence. Extracellular volume is tied to vascular filling, interstitial distribution, pressure relationships, transport capacity, and the broader stability of tissue exchange. Sodium helps define the osmotic character of this compartment, and therefore helps determine whether extracellular space can maintain its volume without excessive contraction or expansion relative to neighboring compartments. The relevance of sodium does not lie only in how much sodium the body contains in total. It lies in where sodium is concentrated and what osmotic force that concentration produces across extracellular space. Whole-system quantity is therefore an incomplete description of extracellular regulation. Placement is the decisive variable.

This is also why sodium is frequently misunderstood when interpreted through overly simplistic hydration language. If sodium is treated merely as a substance associated with water retention, its role appears one-dimensional. But extracellular sodium does not simply "hold water" in a crude sense. It shapes an osmotic field that influences how water is partitioned across vascular, interstitial, and transmembrane contexts. That field interacts with membrane impermeability to certain solutes, renal handling, hormonal regulation, and pressure forces across tissues. Sodium therefore participates in a broad architecture of extracellular stability. To reduce it to a hydration additive or a simple retention factor is to flatten a regulatory mechanism into a consumption model.

Potassium occupies the corresponding intracellular role, though its significance is often obscured by interpretations that treat all fluid as continuous. Potassium is concentrated predominantly inside cells, and that concentration contributes to intracellular osmotic conditions, electrical behavior, and volume maintenance. Just as extracellular sodium helps define the osmotic identity of the extracellular compartment, intracellular potassium helps define the osmotic identity of the intracellular compartment. Water movement into and out of cells is influenced by how intracellular solute concentration relates to extracellular conditions, and potassium is a major part of that intracellular arrangement. Cells fluid content is regulated through the solute structure maintained within them.

This is critical because intracellular balance is not merely a matter of whether cells contain enough fluid. It is a matter of whether intracellular water remains proportionate to the osmotic forces generated by intracellular solutes and to the regulated differences across the cell membrane. Potassium contributes to this balance both by shaping intracellular concentration and by participating in the wider gradient structure that preserves cellular identity. If intracellular potassium distribution changes substantially, the result is not simply a change in mineral status. It is a change in the osmotic and electrical conditions under which the cell operates. Water movement follows those changes. Cell volume, signaling behavior, and functional integrity are therefore linked to potassium not as a general nutrient abstraction, but as an intracellular determinant of compartmental stability.

This intracellular role also clarifies why compartment distinctions must never be collapsed into a single fluid pool. The body does not possess one uniform water reserve whose condition can be described adequately by aggregate volume. It contains intracellular fluid, interstitial fluid, plasma volume, and other organized spaces, each governed by different relationships. Sodium and potassium help define the most fundamental partition between outside and inside. If that distinction is blurred, fluid regulation becomes conceptually distorted. Water entering the system does not automatically become useful in all compartments at once. It distributes according to osmotic structure. A compartment can be relatively expanded while another remains constrained. A cell can gain or lose water not because total fluid has changed dramatically, but because the solute relationships directing transmembrane movement have shifted.

This is the point at which the common assumption that more water necessarily supports fluid balance begins to fail. More water does not automatically resolve distributional instability. If osmotic gradients are not supportive of appropriate partitioning, additional water may alter concentration without restoring correct compartmental relationships. It may dilute one space while failing to correct the distributional forces causing dysfunction in another. It may shift pressure or volume in ways that complicate rather than clarify system behavior. The issue is that water alone does not govern where water ends up. Electrolyte-driven osmotic structure governs that outcome.

For the same reason, fluid imbalance cannot be interpreted as dehydration alone. A system may exhibit reduced effective extracellular volume, altered intracellular hydration, or inappropriate compartmental shifts without the disturbance being well described as simple lack of water. Water may still be present in substantial quantity, but partitioned in ways that reduce functional stability. Conversely, total water reduction may occur alongside relatively preserved compartmental distribution for a time through compensatory mechanisms. In both cases, the more informative question concerns where fluid is moving and why, not simply whether fluid is generally absent.

That "why" returns the analysis to osmotic forces. Water movement between compartments is governed by concentration differences in osmotically active substances, especially electrolytes that are unevenly distributed across membranes and tissue spaces. Those differences create directional pressure for water to move until opposing forces or regulatory controls limit that movement. It is an electrolyte problem, a membrane problem, a compartment problem, and a regulation problem at once. Water behavior cannot be separated from the structures that direct it. Any explanation that begins and ends with fluid intake leaves the underlying mechanism underdeveloped.

This becomes even more apparent under changing conditions. During fasting, heat exposure, stress, or other physiological loads, water distribution may shift because electrolyte handling shifts. Renal conservation patterns may change. Hormonal signals may alter sodium retention or excretion. Intracellular and extracellular relationships may be reweighted under stress. Fluid may move across compartments in response to altered osmotic gradients rather than because water itself has become the primary governing variable. A person may consume water while the more consequential event is the changing arrangement of electrolytes that determines where that water can be held, where it cannot be retained, and which compartments bear the burden of the shift.

This is also why the language of hydration so often becomes misleadingly consumer-oriented. It tends to imply that fluid status is corrected by matching loss with intake in a relatively direct way. It is a distribution problem governed by compartmental gradients, membrane properties, osmotic relationships, and regulatory control. Water entry is only one event within that larger system. Whether that water supports extracellular volume, shifts into cells, leaves the vascular space, or alters concentration relationships depends on the osmotic architecture into which it enters. Without reference to that architecture, the concept of hydration remains materially simplistic.

Even the phrase osmotic balance can be misunderstood if it is treated as a static endpoint. In practice, fluid distribution is dynamic. Water is constantly moving, but its movement remains constrained by compartmental structure and solute distribution. Stability does not mean no movement. It means movement without loss of organized partitioning. Extracellular and intracellular spaces exchange water in ways that preserve relative coherence rather than dissolving their distinctions. Sodium and potassium concentrations are maintained in patterns that allow these exchanges to occur without erasing the identity of the compartments themselves. Fluid order, therefore, is active order. It is maintained through continual regulatory work.

That work includes membrane transport, renal handling, endocrine signaling, and the preservation of concentration differences that prevent fluid from distributing indiscriminately. If those systems become constrained, fluid balance may deteriorate even when water intake appears sufficient in crude terms. Again, the deeper issue is distribution. Water cannot independently decide its own location within physiology. It follows osmotic conditions shaped by electrolyte arrangement, and those conditions are themselves being actively maintained or altered by broader regulatory systems. The path from intake to stability is therefore indirect and contingent. Water enters the system materially, but its physiological significance depends on where electrolyte structure permits it to go.

Hydration is not best understood as the simple presence of water or the replacement of fluid volume in general. It is better understood as the maintenance of appropriate fluid distribution across compartments whose osmotic relationships are governed by electrolytes. Sodium gives the extracellular space much of its osmotic definition. Potassium gives the intracellular space much of its own. Water shifts in response to those structured differences. When those differences remain coordinated, volume, pressure, and compartmental integrity can be maintained. When they become distorted, fluid behavior becomes unstable even if water remains present in substantial amount. The visible subject may appear to be water, but the directing logic is electrolyte distribution.

Concentration vs Total Load

Stability depends on concentration, not total quantity. That is the governing mechanism. Electrolyte behavior becomes physiologically meaningful not when a certain gross amount of material exists somewhere within the system, but when concentration relationships inside specific compartments remain within ranges that preserve signaling, coordination, and functional integrity. Total load is a broad material fact. Concentration is an operational condition. A system may contain substantial total electrolyte quantity and still become unstable if compartmental concentration falls outside the ranges required for electrical behavior, osmotic regulation, and coordinated exchange. For the same reason, a reduced total load does not always produce immediate disruption if concentration is temporarily preserved through redistribution or regulatory compensation. What matters first is not inventory in the abstract, but effective concentration where function is actually being governed.

This distinction is necessary because quantity is easier to imagine than concentration. Quantity suggests substance that can be counted, accumulated, measured, and compared across time. But physiology is not governed by total arithmetic alone. It is governed by compartmental organization. Electrolytes do not operate in a single undifferentiated pool. They exist within extracellular and intracellular environments, within vascular and interstitial spaces, within fluid volumes that can expand, contract, dilute, concentrate, and shift. The same total amount can therefore produce different functional effects depending on how much fluid is present around it, where that fluid is located, how the electrolytes are distributed within it, and what ratios are preserved between one ion and another. Material presence is only the beginning of the analysis.

The difference between total load and effective concentration can be stated more precisely. Total load refers to the overall quantity of an electrolyte present within the organism or entering it over time. Concentration refers to the amount of that electrolyte relative to the volume and compartment in which it is dissolved. A total amount only acquires physiological meaning through its relationship to volume, distribution, and compartmental boundaries. Sodium in extracellular fluid matters according to its concentration within that space. Potassium inside cells matters according to its concentration within the intracellular environment. Magnesium matters according to how it is distributed, buffered, bound, and available within contexts where signaling and enzymatic coordination depend on it. A large total amount that is diluted across expanded volume or displaced into less functionally relevant spaces may fail to preserve the signaling conditions that a smaller but properly concentrated amount could sustain.

This is why total intake or total load can appear sufficient while effective stability deteriorates underneath. The material may be present, yet the concentration relationships that allow the system to operate may already be narrowing. The problem is not merely one of not enough substance. It may instead be one of too much surrounding fluid, altered compartmental partitioning, or disrupted ratios across key electrolytes. If concentration is the relevant operational condition, then total quantity becomes secondary until it is interpreted in relation to the compartmental environment in which it is acting. Physiological stability is never produced by amount alone. It is produced by amount situated correctly within volume and ratio.

Dilution makes this especially clear. Dilution is a mechanism of instability in which concentration falls despite the possibility that total load remains unchanged or even increases. This is difficult to grasp if physiology is interpreted through raw quantity alone, because the presence of material creates the appearance of adequacy. But concentration does not ask only whether an electrolyte exists. It asks how densely that electrolyte is represented within a given volume and compartment. If the surrounding fluid expands disproportionately, the same total amount becomes less concentrated. If water distribution shifts in ways that enlarge one compartment without a corresponding increase in the relevant electrolytes, concentration changes even though total electrolyte quantity may remain materially similar.

This matters because signaling depends on concentration-sensitive thresholds, gradients, and ratios, not just bulk material existence. A diluted extracellular sodium environment alters osmotic behavior and the directional relationships that govern fluid partitioning. A diluted extracellular or intracellular environment can change excitability, transport behavior, and membrane-associated signaling. It is that what is present is no longer concentrated in a way that preserves functional integrity. In this sense, dilution is not a cosmetic shift in numerical density. It is an alteration of regulatory conditions. The system encounters a different operating field because the concentration landscape has changed.

Dilution is often overlooked precisely because it resists the language of deficiency. Deficiency implies lack of total substance, whereas dilution can occur in the presence of substantial or even rising load. More fluid does not necessarily support stability if the added volume changes concentration relationships unfavorably within key compartments. A system can become less stable not because it has lost all material support, but because the material present is spread across a fluid context that no longer sustains the concentrations required for signaling and coordination. When the conceptual frame is limited to total load, dilution becomes almost invisible. The total may appear adequate while the concentration has already moved into unstable territory.

This is also why fluid and electrolyte analysis cannot be separated cleanly. Concentration always implies a relationship between solute and surrounding volume. To describe electrolyte stability without reference to dilution is to omit one of the main ways instability can emerge even in the presence of apparently adequate total load. A compartment expanded by excess water relative to its electrolyte content does not merely contain more fluid. It contains a different signaling condition. The change may alter osmotic gradients, shift transmembrane water movement, distort electrical responsiveness, and increase regulatory strain as the system attempts to preserve function under altered concentration. The relevant event is not simply that more water exists. The relevant event is that concentration has changed.

Depletion operates differently, but it reaches the same terrain of instability by another route. In depletion, effective concentration falls because available electrolyte quantity declines relative to compartmental needs and volume conditions. Yet even here, the deeper issue is not total absence in the abstract. It is the reduction of concentration within compartments where signaling and regulation depend on a maintained range. A depleted system is not unstable merely because some total number has gone down. It is unstable because the concentrations required for gradients, membrane behavior, osmotic balance, and coordinated response are becoming harder to preserve. Depletion becomes functionally relevant at the point where concentration relationships can no longer support the system architecture built upon them.

That distinction matters because depletion is rarely uniform. Electrolytes may fall in one compartment more than another. Extracellular concentration may narrow while intracellular relationships are defended temporarily. Redistribution may preserve one domain while straining another. Hormonal and renal responses may delay visible collapse by narrowing excretion, mobilizing reserves, or altering transport dynamics. Total loss, then, is not always immediately identical to functional loss.

The system may absorb a degree of depletion through compensation. But compensation has limits, and those limits are encountered when concentration can no longer be preserved without excessive strain or when preserving one compartment begins to destabilize another. Again, the decisive variable is concentration, not quantity isolated from context.

This is why imbalance is often misread when it is treated as simple deficiency. Deficiency language assumes that the main question is whether the system has enough material. But imbalance frequently reflects altered concentration relationships before it reflects outright exhaustion of total load. A diluted compartment can behave as though it is lacking what is materially present. A redistributed electrolyte can produce instability despite no dramatic change in system-wide quantity. A depleted compartment may be partly masked by compensatory movement from another space. The language of lack becomes too blunt if it cannot distinguish depletion from dilution, or either of those from redistribution.

Once concentration is recognized as the more relevant operating variable, ratio becomes unavoidable. Electrolytes do not function independently. Their significance lies partly in how they relate to one another within shared spaces and across boundaries. A given concentration of sodium has physiological meaning in relation to water distribution, to counterbalancing ions, and to the preserved asymmetry between extracellular and intracellular environments. A given concentration of potassium matters in relation to membrane behavior, to sodium gradients, and to the conditions under which excitable tissues maintain responsiveness. Magnesium matters not only according to its own concentration, but according to how it participates in stabilizing signaling environments, modulating transport, and influencing the behavior of systems dependent on other ions. Stability therefore depends not only on isolated concentrations, but on concentration ratios that preserve orderly interaction.

Ratio is a more demanding concept than quantity because it forces attention away from one-dimensional sufficiency. One electrolyte may be diluted more than another. One may be depleted while another remains relatively preserved. One may accumulate in a compartment where its increased concentration distorts the balance required for signaling. Under these conditions, the issue is not simply that something has risen or fallen in isolation. The issue is that the relationship between ions has changed, and that relationship often carries more functional significance than any single total amount viewed alone. Signaling is relational. Coordination depends on preserved differentials and proportions, not merely on the presence of individual components in bulk.

This is especially important in electrically active systems. Membrane behavior is shaped not only by how much sodium or potassium exists overall, but by the relationship between their concentrations across the membrane and within the relevant compartments. The same principle extends to osmotic regulation. Fluid distribution is influenced not by isolated total load alone, but by the relative concentration of osmotically active particles in extracellular versus intracellular environments. If one side becomes diluted or one ion becomes disproportionately depleted, the ratio changes and so does the system behavior that follows from it. A ratio disturbance can therefore produce instability without requiring that every total quantity become obviously low. The system can be materially stocked yet relationally disordered.

This relational disorder is often difficult to detect conceptually because broad intake language encourages one-to-one thinking. More electrolyte seems as though it should straightforwardly support greater stability. More fluid seems as though it should straightforwardly support better balance. But concentration and ratio disrupt that simplicity. More of one component can destabilize the relationship with another if volumes, compartments, or gradients are not aligned. More total load can coexist with reduced effective concentration if dilution occurs. More intake can fail to restore intracellular concentration if extracellular distribution dominates the response. The system does not translate gross addition directly into stability. It translates distributed, compartment-specific, ratio-sensitive conditions into stability.

That is why increasing quantity alone does not produce coherence. Stability is not manufactured by accumulation. It is preserved by correct concentration within compartments and correct ratio across interacting ions. Additional quantity may enter the system and still fail to correct instability if the disturbance lies in distribution, dilution, compartment expansion, constrained transport, or disrupted regulatory handling. In some cases, extra quantity may alter total load while leaving the decisive concentration problem unresolved. In other cases, additional fluid may worsen dilution, or additional electrolyte may distort a ratio that was already under strain. Quantity can support stability only when it enters a regulatory environment capable of placing it correctly and preserving the concentration relationships that matter. Without that, amount remains materially real but functionally incomplete.

This is one reason compartment thinking is indispensable. Concentration is never abstract. It always belongs to a place. Extracellular concentration means something different from intracellular concentration. Vascular concentration means something different from interstitial concentration. A total amount distributed across all body water cannot explain what a membrane experiences locally or what an excitable tissue encounters in its immediate environment. Function is determined in compartments, not in the aggregate. Any model that averages the system into a single pool risks losing the very distinctions that give concentration its physiological meaning. Stability belongs to organized spaces. Whole-system totals only become useful when interpreted through those spaces.

The same is true of load. Load can be total, but function is local and relational. A large total load may be compatible with instability if concentration is diluted in the wrong compartment. A modest total load may temporarily sustain stability if ratios and compartmental concentrations are still being defended effectively. This does not make total quantity irrelevant. It makes it subordinate. Total load matters because it places constraints on what concentration can be maintained over time, but it does not determine moment-to-moment functional integrity on its own. Functional integrity depends on whether the system can preserve appropriate concentration relationships where regulation is actually occurring.

This preserved concentration is dynamic rather than static. Volumes change. Water moves. Electrolytes shift between compartments. Renal handling alters retention and excretion. Hormonal systems adjust in response to load, stress, and changing fluid conditions. Concentration therefore cannot be understood as a fixed number detached from process. It is the moving result of regulatory work. Stability does not mean that concentration never changes. It means that concentration remains within a range compatible with signaling and coordination despite continual movement and changing demands. Dilution, depletion, and ratio disruption become destabilizing when they push this moving system outside the ranges it can absorb without loss of functional integrity.

Seen clearly, this changes the meaning of sufficiency. Sufficiency is not the possession of a total amount judged adequate in the abstract. It is the preservation of concentration and ratio within compartments that matter for physiological coordination. A system can appear sufficiently supplied at the level of total load while becoming functionally strained at the level of concentration. Conversely, a temporary reduction in total quantity may remain functionally tolerable if concentration is preserved through distributional control. The deeper question is always the same: not how much exists in gross terms, but whether the concentration relationships that govern gradients, signaling, osmotic balance, and coordinated response remain intact.

This is why instability should not be imagined as a simple line from low quantity to dysfunction. The more accurate line runs from altered concentration to altered system behavior, whether that altered concentration emerges through depletion, dilution, ratio disruption, or redistribution under load. Once that order is reversed and total load is treated as the primary measure of stability, interpretation becomes progressively less precise. Physiology does not operate on gross possession. It operates on concentration within compartments, on preserved ratios between interacting ions, and on the capacity of regulation to keep those relationships from collapsing under changing conditions.

Electrical Stability

Electrolytes maintain predictable electrical behavior. That is the governing mechanism. Electrical signaling in physiology is not created out of nothing at the moment a signal appears. It emerges from a preexisting condition of organized charge separation across membranes, maintained through electrolyte gradients that establish the baseline electrical state from which signaling becomes possible. The relevant question is not simply whether a signal can occur, but whether the conditions beneath that signal are stable enough to make it repeatable, interpretable, and coordinated across time. A system becomes usable not because it can fire once, but because it can fire in a manner that remains structurally reliable under ordinary and changing conditions. Electrolytes are central to that reliability because they maintain the gradients from which membrane potential, threshold behavior, and signaling coherence emerge.

The membrane potential is the first expression of this logic. It is the baseline electrical state created by unequal ion distribution across the cell membrane. Sodium, potassium, and other charged particles are arranged differently inside and outside the cell, and that asymmetry produces an electrical difference across the membrane. The cell at rest is therefore not electrically neutral in the functional sense. It exists in a condition of maintained polarization. That polarization is not passive background. It is the system's stored readiness for controlled signaling. Without it, excitability would not be properly structured, and communication would lose its baseline architecture.

This is why membrane potential must be understood as a maintained condition rather than a static fact. It persists because gradients are continuously preserved in a way that holds the cell in a predictable electrical state between signaling events. The cell membrane is selectively permeable, transport systems continuously defend concentration differences, and the resulting electrical asymmetry is sustained as a functional baseline. If that baseline were not maintained, the system would not merely lose a resting number. It would lose the precondition for orderly excitability. Electrical behavior would no longer emerge from a stable starting point. It would emerge, if at all, from a more variable and less interpretable state.

This distinction matters because electrical signaling is often imagined too narrowly as a series of isolated discharges, impulses, or activations. But a signal is only meaningful relative to the baseline from which it departs. If the baseline is unstable, then the same input may produce different outputs at different moments, not because the input changed, but because the membrane potential beneath it has drifted. A cell that begins from one electrical state is not equivalent to a cell beginning from another, even if the stimulus appears identical. This means signaling is not fundamentally event-based. It is state-dependent. Electrolyte gradients matter because they stabilize the state.

That state-dependence leads directly to threshold behavior. Cells and tissues that signal electrically do not respond continuously to all input with equal intensity. They operate according to thresholds. A threshold is the point at which input becomes sufficient to trigger a defined electrical response. It is a core regulatory feature. Thresholds allow systems to distinguish between background fluctuation and meaningful activation. They make signaling selective rather than indiscriminate. But threshold behavior is only reliable when the underlying electrical conditions are stable. If membrane potential drifts because the gradients maintaining it are altered, threshold does not remain fixed in practical terms. The same input may become insufficient in one state and excessive in another. Predictability begins to erode.

It is a functional relationship between current electrical conditions and incoming stimulation. If the membrane is held consistently within an appropriate electrical range, then threshold behavior becomes reliable. Signals occur when they should occur and remain absent when they should remain absent. But if the underlying electrolyte-driven conditions become unstable, threshold behavior becomes less interpretable. The system may fire too easily, too inconsistently, too slowly, or with variable response to similar inputs. What appears on the surface as altered responsiveness is often a deeper problem of electrical baseline integrity.

This makes electrical stability fundamentally a question of repeatability. A stable system is not simply one that can generate electrical activity. It is one that can generate the same type of activity under the same conditions with a degree of consistency sufficient for coordination. A nerve impulse becomes meaningful because it arises within a framework where membrane conditions, threshold behavior, and conduction properties are sufficiently preserved that the signal can be trusted. A muscle cell becomes functionally usable because its responsiveness is not erratic. A cardiac conduction pathway becomes stable because the electrical events passing through it are not being generated from shifting baseline conditions that introduce drift or unpredictability.

Stable signaling, then, is not merely strong signaling. It is signaling that is temporally and structurally coherent. It occurs with consistent onset relative to threshold, consistent propagation relative to tissue properties, and consistent interpretability relative to surrounding cells. This coherence is only possible because the electrolyte conditions beneath signaling remain within sufficiently narrow ranges. When sodium and potassium gradients remain properly organized, when membrane potential remains appropriately maintained, and when excitability remains proportionate to input, the system can produce signals that mean the same thing over time. It is the quiet continuity that allows signaling to remain legible.

Unstable signaling is different not because it ceases entirely, but because it loses consistency. Signals may still occur, but they become harder for the system to organize around. A stimulus that once produced a clear and predictable response may now produce a weaker, earlier, delayed, excessive, or irregular one. Timing may shift. Responsiveness may fluctuate. Signal propagation may become less uniform. This is not random malfunction. It is the electrical consequence of altered underlying conditions. When electrolyte gradients change, the baseline electrical state changes with them, and signaling behavior becomes correspondingly less stable.

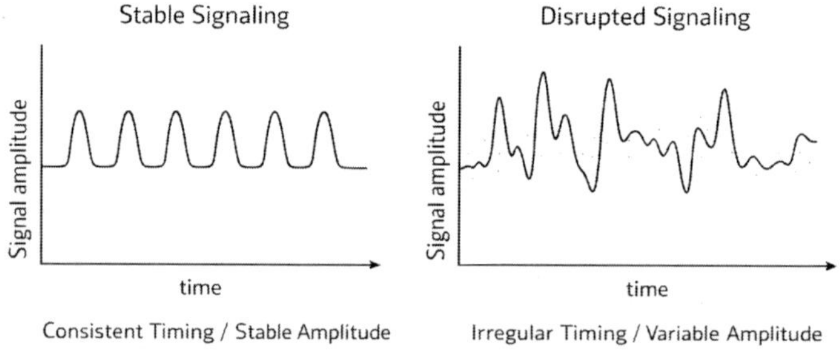

[Figure 3: Stable vs unstable signaling patterns, showing consistent propagation versus irregular, degraded signal behavior under disrupted electrolyte conditions.]

This contrast between stable and unstable signaling is essential because it reveals what electrolytes are doing beneath the surface. They are not simply enabling signals to exist. They are preserving the conditions under which signals remain interpretable. A signal in physiology is useful only if it can be distinguished from noise, repeated when needed, withheld when inappropriate, and propagated in a way that surrounding structures can respond to coherently. Electrolytes help establish that order by maintaining the gradients and membrane conditions that make threshold behavior reliable. When those conditions narrow or distort, the system does not merely lose signal volume. It loses signal quality.

Signal quality depends heavily on consistency because physiological coordination is cumulative. Systems do not respond to one electrical event in isolation. They respond to patterns of activation distributed across time and tissue. If those patterns become noisy, even subtle instability can propagate into larger coordination problems. A tissue receiving inconsistent input cannot respond with the same precision as one receiving stable input. Repeated small deviations in threshold or membrane behavior may not initially appear dramatic, but they alter the reliability of signaling over time. Electrical stability therefore cannot be judged solely by whether obvious failure has occurred. It must be understood in terms of whether baseline conditions still support repeatable and coherent communication.

This is where drift enters the analysis. Drift refers to a gradual deviation in baseline electrical conditions that alters signaling behavior without necessarily producing immediate collapse. A membrane potential that becomes less stable over time does not have to cross an obvious catastrophic threshold to create disruption. It may instead slowly alter how easily cells fire, how rapidly they recover, or how consistently they respond to ordinary inputs. Drift matters because physiology depends on narrow operating ranges. Small shifts in gradients can change excitability enough to affect function long before overt failure becomes visible. The system may still be active, but it becomes less precise.

Drift is often overlooked because it does not present as a discrete event. It is not always dramatic, and it is not always localized to one obvious point of failure. Instead, it changes the background conditions under which many signals are generated. This is why it is better understood as a degradation of signal environment rather than a single signaling defect. A system operating under drift may remain functional in a broad sense while becoming less stable in its timing, less consistent in its responsiveness, and less efficient in its coordination. The issue is not merely that something is happening less well. The issue is that the electrical baseline has become less dependable as a platform for future signaling.

Noise is a related but distinct consequence. Noise refers to unwanted variability or spurious activity within a signaling system that reduces clarity. In a stable electrical environment, the distinction between meaningful signal and background fluctuation remains relatively clear. Thresholds are preserved, membrane potential is held within usable ranges, and activation occurs with a degree of regularity that allows interpretation. But when electrolyte conditions degrade, the separation between signal and noise narrows. The system may become more vulnerable to inappropriate activation, irregular responsiveness, or fluctuating baseline behavior that contaminates otherwise orderly signaling. Often it produces the opposite: activity that is less coherent precisely because it is less constrained.

That loss of constraint is important. Electrical systems are not most stable when they are maximally active. They are most stable when activation occurs under appropriate conditions and remains absent under inappropriate ones. Electrolyte-driven gradients support this selectivity by preserving membrane potential and threshold relationships. When those relationships are altered, selectivity weakens. The system becomes less able to distinguish between sufficient and insufficient input, between relevant and irrelevant perturbation, between orderly activation and background disturbance. What emerges is not merely dysfunction in the abstract, but a more permissive and less interpretable electrical environment.

Signal degradation therefore should not be imagined as a sudden disappearance of function. A tissue may still respond, but with less precision. Signals may still travel, but with less consistency. Coordination may still occur, but with more variability. This degradation matters because physiological systems rely on cumulative trust in repeated signaling. Nerves, muscles, glands, and excitable tissues do not continuously renegotiate what a signal means. They operate under the assumption that the electrical conditions producing those signals are sufficiently stable to make them interpretable. Once that assumption weakens, coordination becomes progressively more difficult.

This difficulty is not confined to the site where the altered signaling first emerges. Electrical instability is inherently prone to propagation because physiology is interconnected. A membrane potential shift in one tissue alters not only local excitability, but the inputs delivered to downstream structures. A threshold change in one part of a signaling chain can alter timing, patterning, and responsiveness elsewhere. Noise in one domain can become uncertainty in another. Drift in one tissue can impose compensatory strain on surrounding systems that must adjust to preserve coherence. Its meaning lies in how instability at one point alters coordination across a larger network.

That networked character is especially important because many physiological systems rely on synchronized or sequential electrical behavior. Timing matters. Order matters. Relative delay matters. Refractory behavior matters. If baseline electrolyte conditions distort any of these through altered membrane potential or threshold responsiveness, the consequences extend beyond the immediately affected cells. A small instability in one node of a network can change the interpretability of incoming or outgoing signals elsewhere. The problem is therefore not simply that one location is behaving abnormally. The problem is that a system built on coordinated signaling has become less coordinated because the electrical language shared across its components is losing consistency.

This makes electrical instability a systems problem even when its first signs appear local. It also explains why the same gross quantity of electrolytes can support different degrees of stability depending on how gradients are maintained. Electrical function is not determined by presence alone. It is determined by whether concentration differences across membranes remain organized enough to preserve membrane potential and threshold behavior within usable ranges. Small changes in these gradients can alter excitability disproportionately because electrical systems are inherently sensitive to baseline conditions.

That sensitivity is not fragility in a simplistic sense. It is the consequence of operating through thresholds and gradients. Systems built to respond quickly and selectively must also remain carefully regulated. Electrical responsiveness is valuable precisely because it is fast and discriminating, but those same features make it dependent on maintained baseline order. Electrolytes provide much of that order by preserving the concentration relationships and membrane asymmetry that allow rapid signaling to remain coherent instead of chaotic. When those relationships narrow or distort, the cost is not only altered activity. It is altered interpretability of activity.

Seen clearly, electrical stability is less about dramatic firing and more about preserved conditions. Membrane potential is the baseline state. Threshold is the gate that determines whether activation becomes meaningful. Consistency is what makes repeated signaling usable. Drift and noise are what emerge when the electrolyte-supported structure beneath those properties begins to erode. Once that erosion occurs, signaling no longer functions as a clean local event. It becomes a source of uncertainty distributed across tissues that depend on repeatable electrical coordination.

Sodium Regulation

Sodium anchors extracellular volume and system pressure. That is the governing mechanism. Sodium is not physiologically important merely because it is present in the organism as one electrolyte among others. Its significance lies in the fact that its concentration and distribution define much of the osmotic structure of the extracellular space, and with that structure come consequences for fluid retention, pressure relationships, transport conditions, and system-wide coordination. It helps determine whether extracellular fluid can remain where it is functionally required, whether vascular and interstitial spaces retain appropriate volume relationships, and whether extracellular conditions remain stable enough to support signaling and exchange across tissues. To understand sodium accurately, it must be treated as a regulator of extracellular architecture rather than as a simple consumable input.

This distinction matters because extracellular fluid is not an incidental container surrounding cells. It is a functional environment through which pressure, circulation, nutrient transport, signaling exchange, and tissue-level coordination are continuously mediated. The extracellular space includes plasma and interstitial fluid, each of which participates in the broader task of maintaining distributional order outside cells. Sodium plays a central role in shaping that order because it is the dominant extracellular cation and therefore one of the primary determinants of extracellular osmotic force. Water distribution follows osmotic structure, and osmotic structure is heavily influenced by sodium. For that reason, sodium is not merely associated with fluid balance. It helps define the extracellular conditions under which fluid balance becomes possible at all.

To say that sodium anchors extracellular volume is to say that extracellular fluid retention depends substantially on sodium being present in the right concentration and managed in the right way across that compartment. It is retained because the extracellular space maintains an osmotic identity strong enough to hold water within its functional boundaries relative to intracellular space and relative to sites of ongoing exchange. Sodium is central to that identity. If extracellular sodium concentration changes, extracellular fluid behavior changes with it. Water shifts in relation to the osmotic environment sodium helps create. This is why extracellular volume cannot be understood as a simple water problem. It is, in large part, a sodium-structured distribution problem.

That structure is what gives sodium its regulatory significance. Sodium is often interpreted too materially, as though its main physiological meaning lies in how much of it has entered or left the system over a defined interval. But sodium's functional importance lies less in raw amount than in how that amount is distributed across extracellular space, how it is retained or excreted relative to surrounding fluid, and how that arrangement influences the system's ability to preserve pressure and coordination. A given total sodium load can support different degrees of extracellular stability depending on whether it is diluted, concentrated, redistributed, or offset by changes in fluid volume. Quantity is therefore not irrelevant, but it is subordinate to arrangement. Sodium becomes physiologically decisive through concentration and location.

This is why sodium's relationship to fluid retention must be understood through distribution rather than crude holding power. Sodium does not simply "retain water" in an undifferentiated sense. It shapes the osmotic field of extracellular space, and water distribution follows that field. If sodium concentration is sufficient to preserve extracellular osmotic structure, water is more likely to remain appropriately partitioned within vascular and interstitial domains. If that structure weakens, extracellular water distribution becomes less stable. The issue is not that sodium physically traps water as if by attachment. The issue is that sodium contributes to the concentration environment that determines where water can remain. Retention is therefore a distributional outcome, not a simplistic binding event.

This distributional role extends directly into pressure. Pressure relationships in physiology are not governed only by mechanical pumping or vessel tone. They also depend on whether the extracellular space, especially the vascular compartment within it, retains enough fluid volume to preserve effective filling and circulation. Sodium contributes to this because it helps define whether extracellular water remains distributed in a way that supports volume where pressure generation depends on it. It can alter the volume context in which circulation, perfusion, and broader tissue coordination occur. Pressure is not independent of distribution. Distribution is not independent of sodium.

This is also why sodium should not be reduced to a single-topic hydration variable. Sodium participates in osmotic structure, fluid partitioning, volume retention, and the preservation of extracellular conditions under which electrical and transport systems remain coherent. A stable extracellular environment is not simply wet enough. It is organized enough. Sodium helps provide that organization. When sodium distribution is appropriate, extracellular space can function as a consistent medium for transport, signaling, and exchange. When sodium regulation becomes unstable, extracellular space becomes less predictable as a regulatory environment. The consequences then extend beyond fluid alone.

Understanding sodium in this way also changes how loss pathways are interpreted. Sweat, urine, and fasting-associated shifts are often treated as though they were straightforward subtraction problems: sodium leaves the system, therefore sodium status declines in a simple linear way.

But physiological loss is rarely just subtraction. It is almost always a change in distribution and regulatory burden at the same time. When sodium is lost through sweat, the significance depends not only on the amount leaving the surface of the body, but on how that loss alters extracellular concentration relative to fluid loss, how renal handling responds, how extracellular volume shifts, and how pressure relationships are subsequently defended. The meaning of the loss lies in what it changes about extracellular structure, not just in the fact that some quantity is gone.

The same is true of urinary sodium handling. It is part of a regulatory process through which extracellular conditions are being adjusted. The kidneys do not simply remove sodium as a matter of elimination. They participate in determining how much sodium remains available to support extracellular osmotic structure and how much is permitted to leave under prevailing hormonal and fluid conditions. Sodium excretion is therefore a statement about system regulation, not only about substance departure. What matters is not merely that sodium has been lost into urine, but what that loss means for extracellular concentration, volume retention, and the stability of the pressure environment being maintained or altered in response.

Fasting introduces another useful illustration because it demonstrates how sodium behavior changes when the system reorganizes under reduced input and altered hormonal conditions. During fasting, sodium dynamics are often discussed as though they simply reflect absence of intake, but that framing is incomplete. The more relevant question is how sodium handling shifts under changing insulin levels, renal regulation, fluid movement, and altered extracellular requirements. The system may excrete sodium differently, redistribute fluid differently, and defend extracellular volume under a different set of constraints than in the fed state. The significance of sodium in fasting is therefore not reducible to the fact that intake is absent. It lies in how extracellular structure is being re-regulated under changing conditions of hormonal signaling and osmotic balance.

This is important because it prevents sodium loss from being misread as a purely linear depletion model. A linear model assumes that sodium leaves, stores diminish, and the resulting condition can be understood by tallying loss alone. But extracellular regulation is more dynamic than that. Sodium may be lost while concentration is partly preserved through concurrent fluid shifts. Conversely, extracellular sodium concentration may become unstable not because total sodium has collapsed, but because water distribution has changed in a way that alters the ratio between sodium and surrounding volume. Loss therefore cannot be interpreted accurately without reference to distribution. The extracellular compartment is not governed by subtraction alone. It is governed by how sodium and water continue to relate under regulatory control.

This brings low sodium into clearer view. Low sodium is often interpreted too narrowly as deficiency, as though the main issue were simply not having enough sodium in total. But the more informative description is instability of extracellular volume and pressure conditions. If extracellular sodium concentration falls meaningfully, the problem is not only one of less sodium present. It is that the osmotic structure supporting extracellular fluid partitioning has weakened. Water distribution may shift. Extracellular volume may become less effectively retained. Pressure relationships may become less stable. Signaling conditions across tissues may become less coherent. What appears superficially as “low sodium” is therefore better understood as a change in extracellular organization.

This reframing matters because it restores sodium to its actual role as a regulator of extracellular conditions. Low sodium is not merely a small-number problem. It is a structural problem in which the extracellular space may no longer maintain the same degree of osmotic definition relative to intracellular space or within its own vascular and interstitial domains. The consequences are not confined to chemistry. They extend into volume behavior, tissue exchange, pressure stability, and the reliability of the extracellular environment as a medium through which the rest of physiology must coordinate. The more accurate question is not simply how low sodium is, but what that low concentration means for the structure of extracellular regulation.

This is also why sodium instability cannot be understood solely through intake narratives. Intake may influence sodium load, but load alone does not determine whether extracellular volume and pressure remain stable. Conversely, low extracellular sodium concentration may emerge in a context where total sodium is not obviously absent, but where dilution, redistribution, or altered excretion have changed the ratio between sodium and extracellular fluid volume. Stability therefore depends on regulation, not direct arithmetic between consumption and status. Sodium is too central to extracellular architecture to be understood through simple input-output thinking.

That architecture is what makes sodium a structural control point. A structural control point is not merely a variable that changes. It is a variable whose management alters the operating conditions of an entire domain. Sodium fits this description because extracellular fluid behavior, pressure relationships, and fluid partitioning depend so heavily on how sodium is handled. Change sodium handling, and the extracellular environment changes with it. Alter the extracellular environment, and transport, signaling, circulation, and tissue coordination all encounter a different baseline. Sodium therefore sits at a level of regulation where local changes can have system-wide effects.

The idea of sodium as a control point also clarifies why its regulation is so tightly integrated with broader physiological systems. The organism cannot allow extracellular sodium handling to drift freely without consequence, because extracellular volume and pressure are too foundational to overall coordination. Sodium retention and excretion therefore occur within a wider network of control that includes renal function, hormonal signaling, fluid shifts, and pressure-sensitive responses. These systems do not regulate sodium in isolation. They regulate sodium because sodium helps determine the extracellular environment in which many other systems must continue to operate. This makes sodium handling less a narrow electrolyte issue and more a foundational aspect of maintaining usable extracellular conditions.

Seen this way, sodium's significance becomes more exact and less distorted by consumer or deficiency language. Sodium is not simply a substance to be replaced, counted, or feared in abstraction. It is one of the primary determinants of extracellular fluid architecture. It shapes how much volume can be retained outside cells, how that volume is partitioned across vascular and interstitial spaces, and how pressure conditions remain coherent enough for broader system function. Loss pathways matter because they alter this architecture. Low sodium matters because it destabilizes it. Regulatory systems matter because they are constantly working to preserve it. Sodium becomes physiologically meaningful at the level where extracellular conditions remain sufficiently organized for the rest of the system to continue coordinating without losing pressure, volume, or functional integrity.

Potassium Balance

Potassium governs intracellular stability and excitability. That is the governing mechanism. Potassium is not physiologically important merely because it is one of the major electrolytes present within the organism, nor because it appears on lists of essential minerals or nutrient categories. Its significance lies in the fact that it is concentrated predominantly inside cells and that this intracellular dominance is one of the central conditions allowing cells to remain electrically organized, osmotically coherent, and functionally ready. Potassium contributes to the gradients that make membrane behavior possible, to the electrical asymmetry that underlies excitability, and to the conditions through which signaling can occur and then return to readiness again. To understand potassium correctly, it must be treated as a determinant of intracellular order rather than as a standalone consumable substance.

This intracellular dominance is not a minor compositional detail. It is one of the most important defining features of cellular organization. The cells internal environment is intentionally different from the extracellular space surrounding them, and potassium is one of the principal reasons that difference exists. The intracellular compartment contains a high concentration of potassium relative to extracellular fluid, while sodium predominates outside the cell. This asymmetry is not incidental. It is a foundational structural arrangement upon which membrane potential, excitability, fluid behavior, and signaling consistency depend. Without that internal potassium dominance, the cell would not retain the same electrical identity or regulatory stability.

This is where potassium must be separated from ordinary nutrient logic. Nutrient language tends to emphasize supply, sufficiency, and material presence in a broad systemic sense. Potassium becomes functionally significant when it is correctly distributed within the intracellular compartment and when that distribution remains appropriately related to extracellular sodium and membrane permeability. In other words, potassium matters less as a stored substance than as part of a maintained gradient architecture. Its meaning lies in placement and relationship, not just in possession.

That relational architecture becomes clearer when potassium is understood alongside sodium. Potassium does not act in isolation. It exists within one of the most important paired gradient systems in physiology: the sodium–potassium relationship across the cell membrane. Potassium is concentrated primarily inside the cell. Sodium is concentrated primarily outside it. These opposing distributions create a structured asymmetry across the membrane, and that asymmetry is central to electrical and osmotic behavior. The membrane is not simply a divider between two fluid spaces. It is the boundary across which this relationship is held, regulated, and selectively used. The sodium–potassium arrangement is therefore not just a descriptive contrast. It is a functional system.

This system is best understood as a directional framework. The high intracellular concentration of potassium creates a tendency for potassium to move outward when membrane conditions allow it. The high extracellular concentration of sodium creates a tendency for sodium to move inward under its own corresponding conditions. They are the directional consequences of maintained concentration differences across a selectively permeable boundary. The membrane becomes electrically meaningful because these opposing forces are held in readiness. Potassium's intracellular dominance is therefore not simply about what is inside the cell. It is about what the cell is prepared to do when channels open, when signals arrive, and when the membrane shifts from resting conditions into active electrical behavior.

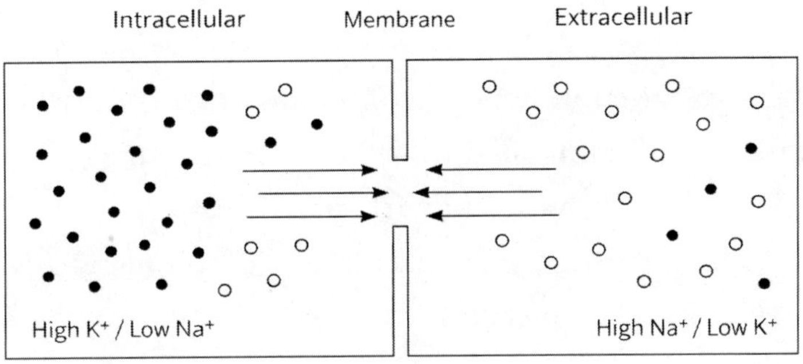

[Figure 4: Sodium–potassium gradients across the cell membrane create directional movement and establish electrical asymmetry and excitability.]

The sodium–potassium relationship is essential because excitability depends on contrast, not on isolated abundance. A cell becomes excitable not because potassium is present inside it in some generic sense, but because potassium is present inside it relative to sodium outside it in a way that creates a stable electrical difference across the membrane. Excitability is therefore a relational property. It emerges from the fact that the inside and outside of the cell are compositionally and electrically different, and that this difference can be altered in a controlled way. Potassium contributes to the resting electrical condition and to the system's ability to repolarize and recover after activation. Sodium contributes to the initiation and directional entry dynamics of activation. The two ions together create a usable electrical cycle. Interpreting potassium alone, without sodium, distorts the mechanism from the beginning.

When potassium is framed as a standalone nutrient, the impression is that its importance can be understood by looking only at its total amount, its intake, or its isolated concentration. But intracellular stability does not arise from potassium in abstraction. It arises from potassium positioned within a coordinated membrane system whose function depends on maintained contrast, selective permeability, and controlled movement. Potassium only acquires its full physiological meaning within this relational architecture. Remove the relationship, and potassium becomes materially present but conceptually incomplete.

This relationship governs excitability because membrane behavior depends on stable baseline conditions before any signal occurs. The cell membrane at rest is not electrically inactive. It is held in a polarized state shaped heavily by potassium distribution and membrane permeability to potassium under resting conditions. Potassium helps define that baseline not as a single momentary value, but as an ongoing condition of electrical readiness. A stable intracellular potassium environment supports stable membrane behavior. An unstable one alters the starting conditions from which all signaling emerges.

That is why excitability should not be interpreted as a property that appears only when cells are actively firing. Excitability is a condition of readiness governed by gradients and membrane state. A cell that is appropriately excitable is not merely able to activate. It is able to activate when appropriate, remain quiet when appropriate, and recover into a state where future activation remains possible without accumulating electrical distortion. Potassium is central to this because it contributes to the maintenance of the resting membrane potential and to the processes by which the membrane returns toward its baseline after disturbance.

Excitability without recovery would quickly become unstable. Potassium participates in both.

Recovery is especially important because electrical signaling is cyclical, not singular. Cells and tissues must activate, reset, and remain capable of activating again with consistency. That reset depends on gradients remaining intact enough to restore membrane behavior after a signaling event. Potassium contributes to that restoration because its movement and distribution help shape the return toward electrical readiness. If potassium gradients narrow or become distorted, recovery becomes less reliable. The issue is not only whether a cell can signal, but whether it can recover into a stable enough state to preserve future signaling fidelity. This is why potassium is not only a signaling ion. It is a recovery ion in the structural sense that it helps preserve repeatable readiness.

This repeatable readiness is what turns electrical behavior into coordinated function rather than isolated activity. A cell that fires unpredictably or recovers inconsistently cannot contribute reliably to a tissue that depends on timing and coherence. Potassium matters because it helps prevent that drift by stabilizing the intracellular conditions on which consistent membrane behavior depends. When potassium gradients are appropriately preserved, excitability remains interpretable. When they are altered, responsiveness can become excessive, blunted, delayed, or inconsistent. What appears on the surface as variable excitability is often a deeper issue of disrupted intracellular electrical structure.

Intake is not irrelevant, but it is too distant from the actual mechanism to serve as the primary explanatory lens. Potassium's role in intracellular stability is not determined directly by how much enters the organism over a short interval. It is determined by whether intracellular distribution remains preserved, whether the sodium–potassium relationship remains coherent, whether membrane transport continues to maintain asymmetry, and whether the gradients supporting excitability and recovery remain intact. A total amount can exist without guaranteeing intracellular stability if distribution is altered, if compartmental relationships shift, or if membrane regulation becomes constrained. Stability is therefore not a direct product of quantity. It is a product of preserved intracellular arrangement.

This is one of the reasons potassium is often misunderstood in simplified physiological narratives. Because it is named as an electrolyte and measured as a substance, it is easy to assume its function is best understood through substance logic alone. But intracellular stability is not governed by substance in the abstract. It is governed by gradients. A potassium ion outside the cell does not have the same physiological meaning as a potassium ion inside the cell. A total potassium load distributed in the wrong way does not preserve excitability. What matters is not just that potassium exists, but that it remains concentrated in the intracellular space in a relationship to extracellular sodium and membrane behavior that sustains electrical asymmetry. The same material becomes functionally different when its placement changes.

A disrupted potassium state is not always best understood as simple absence. The more accurate question is whether the relationship between potassium, sodium, and membrane conditions remains stable enough to support coordinated signaling and recovery. Potassium-related instability may arise through altered distribution, disrupted gradients, dilution, shifts across compartments, or failure of membrane regulation to preserve the normal intracellular-extracellular asymmetry. In these cases, the problem is not merely that there is less potassium in total. The problem is that the relational architecture on which excitability depends has been altered.

This relational instability is often experienced physiologically as unpredictability rather than straightforward silence. A cell or tissue may still signal, but with less consistency. Responsiveness may drift. Recovery may become less reliable. Threshold behavior may change because the membrane no longer begins from the same baseline state. These are not random fluctuations. They are the expected consequences of altered gradients in a system that depends on narrow electrical conditions for orderly behavior. When potassium's intracellular role is understood correctly, excitability changes stop appearing arbitrary. They become interpretable as the outcome of structural shifts in membrane-supported asymmetry.

The sodium–potassium relationship is therefore not merely one component of intracellular physiology. It is one of the central organizing systems through which the cell remains both stable and responsive. Sodium helps define the extracellular environment. Potassium helps define the intracellular environment. The membrane between them preserves the difference. Transport systems maintain the asymmetry. Signaling events selectively use the gradients created by that asymmetry. Recovery restores the membrane toward its baseline state. Potassium's role cannot be isolated from this structure without becoming conceptually distorted. It matters not as a lone actor, but as one pole of a coordinated system.

That coordination extends beyond individual cells because tissues are only as reliable as the membrane behavior of the cells composing them. If intracellular potassium stability degrades across many cells, tissue-level signaling becomes less coherent. Timing changes, responsiveness varies, and recovery patterns become less synchronized. The disturbance is therefore not confined to a microscopic event. It can propagate upward into broader patterns of instability because excitability and recovery are not private cellular matters. They are the basis upon which larger physiological coordination depends. Potassium's intracellular role is therefore local in placement but system-relevant in consequence.

Seen clearly, potassium balance is not best understood as a matter of having enough potassium in general. It is better understood as preserving intracellular dominance within a membrane-regulated, sodium-linked gradient system that sustains electrical readiness, responsiveness, and recovery. The decisive issue is not raw amount alone, but whether the gradients remain intact enough to preserve excitability without noise, activation without instability, and recovery without drift. Potassium becomes physiologically meaningful where intracellular order remains strong enough to keep membrane behavior coherent across repeated cycles of signaling.

Magnesium Modulation

Magnesium regulates system activation and control. That is the governing mechanism. Magnesium is often described in broad and imprecise terms, as though its significance lies in being generally helpful, generally supportive, or broadly associated with calm, recovery, or cellular function. Magnesium is better understood as a modulatory regulator: an ion whose importance lies not primarily in driving large compartmental gradients in the way sodium and potassium do, but in shaping how activation is initiated, constrained, buffered, and returned toward controlled signaling states. Its role is less architectural in the osmotic sense and more regulatory in the dynamic sense. It does not merely participate in physiology. It moderates the conditions under which physiology remains coherent rather than excessive, noisy, or unstable.

This modulatory role is especially visible in the nervous system, where signaling must remain both active and controlled. A functioning nervous system is not one in which signals are maximally present. It is one in which signals occur with enough intensity to carry information, enough selectivity to remain meaningful, and enough restraint to prevent escalation into indiscriminate activation. Magnesium contributes to this restraint not by suppressing signaling into passivity, but by influencing the conditions under which signaling becomes more or less likely to proceed. It shapes excitability through regulation rather than through raw force. Magnesium helps make the systems governable.

That governability matters because nervous system signaling is inherently vulnerable to excess if modulation is weak. Excitable tissues operate through thresholds, channel behavior, synaptic transmission, membrane responsiveness, and repeated cycles of activation and recovery. A system built on these properties must remain sensitive enough to respond, but not so permissive that it amplifies perturbation into instability. Magnesium participates in this balance by influencing channel function, receptor behavior, enzymatic processes, and the broader signaling environment in which activation occurs. Its role is not always visually dramatic, which is part of why it is so often misunderstood. But quiet regulation is not the same as minor importance. In many cases, the less visible the modulatory function, the more foundational it is to signal quality.

This is why magnesium should not be interpreted as a general supportive substance. That language is too vague to preserve mechanism. A supportive substance sounds like a passive addition that assists the system in broad, nonspecific ways. Magnesium does not operate meaningfully in that register. It helps define whether activation is regulated, whether signaling remains appropriately gated, and whether the system can tolerate ongoing excitatory demand without drifting toward electrical or functional instability. These are not background benefits. They are active control functions. Magnesium becomes important where the system must distinguish between activation that is usable and activation that is excessive.

Activation is not inherently destabilizing. It is necessary. Neurons must fire. Muscles must contract. Signaling pathways must open, close, and respond to input. The problem is not activation itself. The problem is activation that loses modulation. When signaling becomes too permissive, too easily amplified, too poorly buffered, or too difficult to terminate cleanly, the system becomes less coordinated even if activity remains high. Magnesium helps preserve the difference between responsive function and uncontrolled escalation. Its role is therefore not to oppose activation as a concept, but to help determine whether activation remains organized enough to be useful.

This is where the balance between activation and relaxation is often misdescribed. Relaxation is usually not the absence of signaling. It is the result of signaling being regulated appropriately enough to permit release, reset, and controlled downshift after activation. Magnesium contributes to this not by acting as a passive sedative force, but by participating in the regulatory conditions that prevent activation from remaining unnecessarily prolonged or excessively amplified. In that sense, what appears outwardly as relaxation is often better understood inwardly as successful modulation. The system is not inactive. It is governed.

That governing function depends on thresholds. Systems become unstable not only when they cannot activate, but when they activate too easily, recover too poorly, or remain too close to the edge of activation without sufficient control. Magnesium influences this terrain by shaping how excitable systems behave around threshold. It affects whether input remains proportionate to output, whether channels and receptors respond within controlled ranges, and whether signaling events terminate with enough precision to restore readiness without residue of excessive activation. Magnesium's role is more distributed and conditional than that. But across many domains, it contributes to the difference between a system that responds cleanly and one that accumulates noise.

This is one of the reasons magnesium is so difficult to understand if approached through a purely material or intake-based frame. The presence of magnesium in the system does not automatically explain whether modulation is functionally effective. Total presence and effective control are not identical. Magnesium may be materially present while its regulatory role is functionally constrained by distribution, binding state, intracellular handling, competing system demands, or broader signaling conditions. It is not enough for magnesium to exist in the organism in gross terms. Its significance lies in whether it is participating effectively in the modulation of activation where control is required.

That distinction matters because instability is often interpreted too narrowly as absence. If signaling becomes excessive, irregular, or difficult to regulate, the reflexive interpretation is often that something is "low" or "missing" in a direct and linear sense. But with magnesium, the more accurate framing is often loss of modulation rather than mere lack of substance. A system can become less buffered, less gated, and less controlled without that necessarily being reducible to a simplistic deficiency model. Magnesium's relevance therefore lies as much in preserving modulation under stress as in being present under ideal conditions.

This is especially important in the nervous system, where excitation must remain interpretable. Signaling that is too sparse fails. But signaling that is too noisy also fails. The nervous system depends on selective amplification, not indiscriminate activation. Magnesium contributes to that selectivity by helping preserve conditions under which channels and receptors do not respond too freely or remain open too easily under ordinary stimulation. It is a matter of controlled permission. The system must allow signaling, but not dissolve into it. Magnesium participates in the regulatory architecture that keeps that distinction intact.

The same logic extends beyond the nervous system because modulation is not only a neural problem. Muscular behavior, vascular tone, cellular signaling pathways, and broader physiological coordination all depend on activation being constrained by regulatory context. Magnesium participates in this broader buffering function because it influences how excitable or responsive systems behave under ongoing demand. A tissue that cannot modulate well may still appear active, but its activity becomes less stable. It may contract too readily, recover less efficiently, or transmit more variability through its signaling patterns. What emerges is not necessarily collapse, but a more difficult-to-govern operating state.

Magnesium matters because it helps reduce the likelihood that ordinary activation becomes cumulative instability.

This is where buffering becomes the most useful conceptual frame. A buffer is not a passive reserve sitting unused until catastrophe. It is a stabilizing influence that absorbs variability, limits escalation, and preserves coherence under changing conditions. Magnesium functions in this way across multiple systems. It helps prevent activation from becoming too abrupt, too prolonged, or too poorly regulated. It limits the extent to which fluctuation becomes destabilizing. This makes magnesium less a driver of large visible movements and more a governor of how those movements unfold. Its role is not to create every signal, but to influence whether signals remain proportionate and recoverable.

This buffering role is particularly important because physiological systems are never operating in isolation from load. Stress, sleep disruption, fasting, illness, environmental pressure, and repeated signaling demand all increase the burden placed on regulatory control. Magnesium's significance often becomes more visible here, not because it suddenly becomes relevant only under strain, but because strain reveals the importance of whatever has been quietly maintaining control all along. A buffered system can absorb more variability without crossing into noise. A poorly buffered one becomes progressively more difficult to stabilize.

That does not mean magnesium should be mythologized as a universal stabilizer or treated as the singular answer to all forms of overactivation. Magnesium is important because it participates in regulation, not because it overrides the rest of physiology. It functions within a coordinated network that includes sodium and potassium gradients, membrane behavior, endocrine signaling, nervous system state, and broader cellular control mechanisms. Its modulatory role only makes sense inside that wider context. Magnesium is not a substitute for regulation. It is one of the means through which regulation remains possible.

This wider context is important because it prevents magnesium from being interpreted as a mood-state mineral, a generalized calming substance, or a vague symbol of restoration. What appears outwardly as calm, control, or reduced activation often reflects the fact that signaling has remained better regulated, thresholds have remained more appropriately governed, and the system has not drifted as easily into unbuffered excitation. Magnesium participates in these outcomes not as a psychological metaphor, but as a physiological modulator of signal behavior. Its role is concrete, but not crude. It is precise, but distributed.

That distributed precision is what makes magnesium difficult to narrate in simplistic terms. It does not dominate extracellular osmotic volume the way sodium does. It does not define the intracellular electrical pole the way potassium does. Instead, it shapes how many systems behave around the edge of activation. It influences whether signaling becomes excessive, whether thresholds remain governable, and whether the system retains enough buffering to keep ordinary activation from becoming cumulative instability. It is a central form of control. Many systems do not fail because activation disappears. They fail because activation loses regulation. Magnesium matters because it helps preserve the conditions under which activation can remain coordinated rather than self-amplifying.

Seen clearly, magnesium modulation is not a soft or optional layer added onto a fundamentally self-regulating system. It is part of the system's capacity to remain governable in the first place. Nervous system behavior, excitability, signaling thresholds, and broader patterns of activation all depend on there being enough buffering, gating, and regulatory control to prevent drift toward excess. Magnesium contributes to that control not through blunt opposition, but through modulation. It helps keep signals usable. It helps keep activation proportionate. It helps keep recovery possible. Its importance lies not in being a general support category, but in being one of the ions through which physiological systems remain capable of activating without losing control of themselves.

Signal Transmission

Electrolytes enable coordinated signal propagation. That is the governing mechanism. Physiological signaling is not best understood as a collection of isolated electrical events appearing independently in separate tissues. It is better understood as sequential transmission: activity that begins under specific conditions, moves through structured pathways, and remains meaningful only if its propagation retains timing, directionality, and interpretability across connected cells and tissues. A signal is not complete at the moment it is initiated. It becomes physiologically relevant only if it can travel, be received, be translated, and participate in a larger chain of coordination. Without those conditions, activation may still occur in fragments, but transmission loses coherence.

This distinction between isolated events and propagated sequences matters because many physiological descriptions stop too early. A single electrical event has limited value if it cannot extend into the surrounding architecture that depends on it. Signaling in living systems is not merely about initiation. It is about continuity. The system requires events to unfold in sequence rather than remain trapped as disconnected points of activity. Transmission is therefore not an optional extension of signaling. It is one of signaling's defining properties.

To transmit a signal is to preserve its legibility as it moves. This requires more than energy or activation in the abstract. It requires maintained gradients, stable membrane conditions, predictable threshold behavior, and sufficient electrical consistency across adjacent structures for each segment of the pathway to respond appropriately to the one before it. It is a chain of orderly depolarization and recovery occurring under conditions that remain coherent enough for sequential communication. Each point in the chain depends on the point before it, but it also depends on its own local electrical readiness. This is where electrolyte gradients become decisive. They are not only important for starting a signal. They are important for allowing the next segment to receive and continue it without distortion.

The signal moves because each part of the pathway remains electrically prepared to translate incoming activation into outgoing activation with sufficient fidelity. That preparation depends on gradients. If membrane potential is stable, if threshold remains appropriately positioned, and if excitability is preserved within usable bounds, then the signal can continue as an ordered sequence. If these conditions deteriorate, transmission becomes vulnerable. The signal may slow, weaken, fragment, or fail to propagate entirely. What changes is not only whether activation occurs at one point, but whether the conditions needed for continuation still exist at the next.

This is why transmission depends so heavily on underlying electrical conditions rather than on the signal event alone. A propagated sequence is always moving through tissue that is already in some electrical state before the signal arrives. That baseline state matters because it determines whether incoming activation will be received, amplified, dampened, delayed, or blocked. Electrolyte gradients shape this baseline by preserving membrane potential, threshold relationships, and the broader readiness of cells to participate in coordinated signaling. It must be explained by the hidden continuity of conditions beneath that movement.

This electrical terrain is often underestimated because the signal itself is more visible conceptually than the state that allows it to move. Signals may continue to pass for a time even as the margin of transmission decreases. Threshold may become more variable. Recovery may become less complete. Membrane readiness may become less consistent across repeated cycles. Under these conditions, the pathway may still function in broad terms, but it becomes more vulnerable to interruption or distortion. Transmission does not fail only when it disappears. It also fails when it loses reliability.

The meaning of a signal often lies in when it arrives relative to other signals, how consistently it propagates, and whether it reaches downstream structures with enough integrity to preserve coordination. A signal arriving too early, too late, too variably, or too incompletely is not functionally equivalent to one that propagates cleanly. This is why propagation must be evaluated in terms of order rather than mere occurrence. The system does not only need signals to happen. It needs them to happen in a sequence that remains interpretable across time and tissue. Electrolytes support this order because they preserve the gradients and electrical consistency that keep transmission from dissolving into noise or fragmentation.

Instability begins when those supporting conditions weaken. In reality, degradation often emerges more gradually. A propagated signal may begin to lose speed, clarity, or consistency before it fails completely. Local membrane conditions may drift. Thresholds may become less stable. Recovery between activations may become less complete. Small alterations in gradients may change how readily a segment of tissue responds to incoming activation. They are structurally driven consequences of altered electrical conditions. The pathway becomes less dependable because the terrain beneath the signal has become less organized.

Signal transmission does not usually fragment because physiology becomes mysteriously unreliable. It fragments because the conditions that once supported sequential propagation have changed. A degraded gradient, a narrowed electrical margin, a less stable membrane potential, or a more variable recovery state can all make transmission less coherent. The signal itself may still be generated at one point, but its ability to move cleanly through the system depends on whether adjacent structures remain prepared to receive and continue it. Breakdown is therefore not just failure of a signal. It is failure of a pathway to preserve the conditions required for orderly continuation.

This pathway dependence is what makes transmission especially vulnerable to instability that propagates through connected systems. A local change in electrical readiness does not remain local if the tissue in question participates in a larger signaling chain. An altered segment can distort timing, weaken continuity, or impose variability on downstream structures that were otherwise capable of functioning normally. It does not stay confined to the original point of degradation. Because signaling is sequential, disruption at one point alters what is delivered to the next, and that altered delivery may then change the behavior of the next structure in turn. The system becomes less coherent not because every part fails independently, but because a connected sequence begins to lose continuity.

This is why coordinated propagation must be distinguished carefully from fragmented signaling. Coordinated propagation is not simply the presence of movement. It is movement that remains ordered across the entire signaling path. Each segment activates in relation to the previous one, each recovery phase preserves readiness for the next cycle, and the signal arrives with enough consistency to retain its meaning. It may involve partial transmission, irregular progression, local bursts of activity that do not integrate cleanly into the broader sequence, or signals that degrade as they travel. Fragmentation is therefore not defined only by absence. It is defined by loss of continuity.

That loss of continuity has important implications because physiological systems depend on functional coherence, not just local excitability. A cell that can still activate is not sufficient if its activation no longer contributes reliably to the sequence around it. A tissue that remains electrically responsive is not necessarily stable if its signaling becomes temporally inconsistent or incompletely propagated. Functional coherence requires that local events remain integrated into broader patterns of communication. Electrolyte-supported gradients help preserve that integration by maintaining the underlying electrical conditions that allow one activation to become the next without accumulating excessive distortion. When those conditions weaken, the system may still appear active while becoming progressively less coherent.

Fragmentation also helps explain why signaling issues are often misread when interpreted only at the point of visible dysfunction. If a downstream tissue receives inconsistent input, the surface appearance may suggest that the downstream tissue itself is the primary problem. But often the more relevant issue lies upstream or within the transmission path connecting them. Because signals are propagated sequences, any point along the route can alter the quality of what is delivered further along. This means signaling dysfunction should not be interpreted solely as a problem of the site where symptoms or visible effects appear. It must be understood as a problem of communication integrity across the pathway. The visible endpoint is often only the final expression of a longer chain of transmission instability.

Gradients are not only what allow local activation to occur. They are what make sequential responsiveness possible across an extended structure. Each part of a signaling pathway must maintain enough membrane asymmetry, threshold stability, and electrical readiness to respond to the signal arriving from the previous part. If gradients collapse or become inconsistent across segments of the pathway, propagation becomes vulnerable to delay, weakening, or discontinuity. The signal then encounters a less reliable medium. Transmission becomes less like a coordinated wave and more like a set of uneven electrical attempts, some of which continue and some of which fail.

That shift from wave to fragmentation is one of the clearest markers of degraded signaling integrity. In a coherent system, propagation is directional, sequential, and interpretable. In a degraded one, activation may still occur, but the pattern becomes less unified. Timing drifts. Some regions respond while others hesitate. Recovery states become less synchronized. The result is not simply less signaling. It is signaling that is less useful because it no longer carries the same degree of ordered communication across the tissue or system. Physiological function depends not on isolated electrical capacity, but on the ability to preserve signal structure as it moves.

This preservation of structure is what makes transmission a systems problem rather than merely a local electrical one. Communication integrity across physiology depends on signals moving consistently enough that tissues can coordinate around them. If transmission becomes unstable, communication degrades even when individual components retain partial function. A system built on sequential signaling cannot remain fully coherent if the sequences connecting its parts become inconsistent. The consequences then extend beyond one tissue or one event. They alter timing relationships, coordination patterns, and the system's ability to behave as a unified whole. Transmission instability therefore belongs to the level of system behavior, not just local excitability.

It is the real operational form of signaling in a connected organism. A signal that cannot move coherently is not yet communication. Electrolytes matter because they preserve the gradients, thresholds, and electrical readiness required for signals to propagate with enough consistency to remain meaningful. When those conditions are maintained, signaling becomes sequential, coordinated, and functionally coherent. When they degrade, propagation becomes fragmented, distorted, or incomplete, and the system loses not just isolated signals, but the continuity of communication on which broader physiological coordination depends.

Contraction and Relaxation

Activation and release cycles depend on electrolyte balance. But in physiology, they are better understood as coordinated phases within a repeating signaling cycle. Contraction is not simply shortening. Relaxation is not simply stopping. Each phase depends on specific electrical and biochemical conditions that must arise in sequence, resolve appropriately, and return the system to readiness for the next cycle. The relevant unit is not contraction alone. It is the cycle.

This distinction matters because muscle behavior is frequently described too mechanically, as though the visible movement tells the whole story. It does not. The visible act of contraction is the surface expression of prior electrical signaling, membrane activation, ion movement, and regulated intracellular events that make contraction possible in the first place. A muscle does not contract because it possesses tissue capable of shortening in the abstract. It contracts because signaling has altered the conditions within the cell in a way that permits activation to proceed. This makes contraction fundamentally a signaling event before it is a mechanical one. The mechanics are real, but they are downstream. If the signaling conditions are unstable, the contraction phase becomes less reliable regardless of how intact the tissue may appear structurally.

To define contraction properly, then, it must be understood as the organized expression of activation. Electrical signaling reaches the muscle membrane, membrane behavior changes, intracellular conditions shift, and the machinery of force generation becomes available. None of this occurs independently of electrolyte-driven gradients. Sodium and potassium contribute to the electrical readiness and propagation of activation signals. Magnesium participates in modulation and control of excitability and signaling behavior. It is a regulated conversion from readiness into action, and that conversion depends on preserved signaling architecture.

This architecture matters because contraction must begin cleanly. A stable contraction event is not simply one in which force appears. It is one in which activation occurs at the appropriate threshold, propagates in a coordinated way, and engages tissue with sufficient timing and coherence that the resulting contraction remains interpretable rather than erratic. Contraction is only useful when it emerges under controlled conditions and remains part of a larger coordinated sequence. A system that contracts unpredictably, incompletely, or without clean initiation is already showing that the cycle beneath the visible movement has begun to degrade.

Relaxation is not what happens when contraction simply ends on its own. It is an actively regulated release phase in which the conditions that sustained activation must be reversed, withdrawn, buffered, or re-stabilized so that force generation can disengage and the tissue can return toward readiness. Release is another form of control. If contraction is the regulated organization of activation, relaxation is the regulated dissolution of that activation. Without active release, the cycle does not complete. It remains partially unresolved.

They remain functional by being able to activate and then recover from activation cleanly enough to activate again. Relaxation is therefore not secondary to contraction. It is coequal in importance. A muscle that can generate force but cannot release efficiently is not stable. A signaling system that can initiate activation but cannot restore baseline conditions is not reliable. Release is what prevents activation from becoming cumulative distortion. It is the phase that restores the system's capacity to re-enter future cycles without carrying unresolved activation forward into the next one.

Electrolytes participate in both phases of this cycle because the transition into contraction and the transition out of it are each governed by changes in signaling conditions, membrane behavior, and intracellular regulation. Sodium and potassium help establish and restore the electrical states required for activation and recovery. Their gradients determine membrane readiness, excitability, and the ability of activation signals to propagate predictably. Magnesium helps modulate the signaling environment, influencing how easily activation escalates and how well the system remains governable during and after active states. None of these ions belongs only to the "on" phase or only to the "off" phase. Their importance lies in how they support the integrity of the transition between phases.

That transition is where stability actually resides. A system is not stable simply because it can contract strongly or relax eventually. It is stable when it can move between activation and release without accumulating distortion. Contraction must begin when appropriate, proceed coherently, and terminate in a way that permits clean release. Relaxation must not merely arrive after delay or exhaustion. It must be actively achieved through restoration of conditions that make future activation possible again. Electrolyte balance matters because it preserves the signaling and regulatory environment in which these transitions remain smooth enough to repeat without progressive degradation.

A single contraction event always occurs in the context of prior cycles and in anticipation of future ones. The system is never merely contracting or relaxing in the abstract. It is continually passing through repeated phases of activation, release, readiness, and reactivation. If any one phase becomes incomplete, the consequences spill forward into the next cycle. An unresolved contraction state alters relaxation. Incomplete relaxation alters the next activation. Distorted activation alters the efficiency and coherence of subsequent force generation. Stability is therefore not a property of any one phase in isolation. It is a property of the continuity linking them.

This is why stiffness and fatigue should not be interpreted too narrowly as local outcomes detached from signaling cycles. Both can reflect disruption in the system's ability to transition cleanly between activation and release. Stiffness is not always best understood as a tissue problem in the structural sense alone. It may reflect incomplete release, residual activation, altered excitability, or impaired restoration of baseline conditions between cycles. A system that does not fully release remains partly engaged, and that partial engagement changes the mechanical and signaling environment in which future movement must occur. The visible experience is stiffness, but the deeper issue may be that the cycle is not resolving cleanly.

Fatigue can be understood through a similar lens. A fatigued system is not merely one that has produced too much output. It may be one in which activation becomes less efficient, propagation becomes less coherent, release becomes less complete, or recovery between cycles becomes progressively less stable. The visible outcome is reduced functional integrity, but the mechanism often lies in a cycle that is becoming harder to preserve under repeated demand.

This is important because it prevents stiffness and fatigue from being misread as isolated endpoints. They are often expressions of a system that has become less capable of moving cleanly between phases. A contraction that begins too easily, ends too slowly, recovers too incompletely, or repeats too inconsistently is already showing instability at the level of cyclical regulation. Electrolyte imbalance can contribute to this not because it directly causes every visible symptom in a simple linear way, but because it alters the signaling conditions on which the cycle depends. If excitability drifts, if membrane behavior becomes less stable, if modulation weakens, or if recovery conditions narrow, the cycle becomes harder to govern.

A system can become stuck in activation not necessarily by remaining maximally contracted at all times, but by failing to release fully enough between cycles. It can remain electrically or mechanically closer to activation than intended, carrying forward residual tension, altered threshold behavior, or incomplete restoration of readiness. Conversely, a system can also become functionally sluggish or delayed if activation becomes harder to initiate cleanly because the baseline conditions required for excitability have drifted too far from useful responsiveness. In both cases, the problem is not simply too much or too little activity. It is loss of cyclical control.

This loss of control often appears as inconsistency. The system may still contract, but not with the same precision. It may still release, but not with the same completeness. Repeated cycles may begin to show variable timing, reduced fluidity, or altered responsiveness under similar demands. They are what should be expected when the electrical and regulatory conditions supporting activation and release become less stable. The muscle is not merely behaving unpredictably for no reason. The signaling environment beneath its behavior has changed. Electrolyte-dependent gradients and modulatory influences help determine whether each cycle begins from a clean baseline or from a progressively more distorted one.

That baseline matters because cyclical systems rely on return as much as on output. A muscle or excitable tissue does not remain usable because it can generate force repeatedly by brute persistence. It remains usable because each cycle includes sufficient release and restoration to preserve future responsiveness. Recovery is built into every successful cycle. It is not what happens after the cycle matters. It is part of what makes the cycle matter at all. Electrolytes contribute to this because they help preserve the gradients, thresholds, and modulatory conditions that allow recovery to occur within the cycle rather than only after breakdown.

This is also why contraction and relaxation should be understood as expressions of signaling quality rather than only of tissue capability. A tissue may remain anatomically intact while its activation-release cycle becomes less stable. If signaling becomes too noisy, too easily triggered, too weakly buffered, or too poorly resolved, the visible mechanical behavior changes accordingly. The issue is not only whether the tissue can still move. It is whether it can still cycle cleanly. Mechanical integrity without signaling integrity does not preserve stable function. Electrolytes matter because they help bridge that gap, maintaining the conditions under which electrical events can become coordinated mechanical phases and then return to readiness again.

Seen clearly, contraction and relaxation are not separate phenomena joined loosely by convenience. They are phases of one controlled cycle. Activation initiates force. Release restores readiness. Electrolytes participate in both by preserving the signaling conditions required for each transition to occur without excessive residue, distortion, or delay. When those conditions remain stable, the cycle remains repeatable, efficient, and coordinated. When they degrade, the system may still produce force or still eventually release, but the continuity between phases begins to fray. What emerges then is not simply weakness or tightness in isolation, but a broader loss of cyclical integrity in which activation and release no longer relate to one another with the same degree of control.

Rhythm Stability

Stable systems maintain predictable oscillatory patterns. That is the governing mechanism. Physiological stability is not expressed primarily as stillness, constancy, or the absence of change. It is expressed through regulated variation over time. Systems remain stable not because nothing moves, fluctuates, or responds, but because those fluctuations occur within organized ranges and in repeatable temporal relationships. Rhythm is therefore not an accessory to physiology. It is one of its most visible and structurally important forms. Electrolyte balance matters within this framework because it helps preserve the gradients, signaling conditions, and coordination required for oscillatory behavior to remain interpretable rather than erratic. A stable system is not a static one. It is one whose movement retains pattern.

This is an important correction because physiology is often misread through event-based thinking. In that frame, the main unit of interpretation becomes the individual event: a contraction, a signal, a shift in fluid distribution, a moment of activation, a measurable change at a single point in time. These events are real, but they are rarely meaningful in isolation. Their significance depends on how they relate to what came before, what follows after, and whether they remain consistent with the broader temporal architecture of the system in which they occur. A heartbeat is not just one beat. A nerve impulse is not just one discharge. A contraction is not just one shortening event. Each belongs to a rhythm, and the rhythm is often more informative than the event itself.

A rhythm is not merely recurrence. It is recurrence with enough structure that the system can anticipate, coordinate, and remain coherent across time. Predictability is central here, but predictability does not mean mechanical exactness. Physiological rhythms are not metronomic in the rigid sense. They allow variability, but that variability occurs within a recognizable and governed envelope. A stable rhythm can speed, slow, deepen, flatten, or adapt within limits while remaining interpretable as part of one coherent pattern. Stability therefore lies not in fixed repetition, but in bounded oscillation.

This bounded oscillation is one of the clearest expressions of system health because living systems depend on alternation rather than permanent occupancy of one state. Activation alternates with release. Excitability alternates with recovery. Distribution shifts and then rebalances. Signaling rises and falls. Pressure changes and resolves. Electrical readiness is used and restored. These are not signs of instability. They are signs that the system remains capable of moving through time without becoming trapped, fragmented, or directionless. Rhythm is the temporal expression of regulation. If regulation is preserved, rhythm remains coherent. If regulation weakens, rhythm becomes more vulnerable to distortion.

Electrolytes matter in this context because oscillatory behavior depends on repeatable underlying conditions. A rhythm can only remain stable if the system repeatedly returns to states from which future cycles remain possible. This return depends on gradients, membrane behavior, threshold stability, fluid distribution, and signaling consistency—all of which are shaped by electrolyte conditions. Sodium and potassium gradients help preserve the electrical architecture that allows excitable systems to activate and then reset. Magnesium helps modulate activation so that oscillations do not become excessively amplified or poorly resolved. Electrolytes therefore support rhythm not by creating "rhythm" as an abstract feature, but by preserving the conditions that allow cyclical behavior to repeat without cumulative distortion.

A single event can appear normal while the pattern containing it is already degrading. This is one reason event-based interpretation can be misleading. If the analysis stops at whether one contraction occurred, one signal propagated, or one fluid shift resolved, it may miss whether the sequence across time remains stable. Rhythm reveals what the event alone cannot: whether the system can continue to behave coherently across repeated cycles without losing timing, range, or internal coordination. Electrolyte-dependent stability often becomes more visible at this pattern level than at the level of any one isolated event.

This is where stable oscillation must be distinguished from unstable variation. Not all variability is adaptive, and not all irregularity is benign simply because living systems are dynamic. Stable oscillations include variation, but that variation remains bounded and organized. The pattern may not be perfectly uniform, yet it retains enough consistency that the system remains predictable to itself. One cycle resembles the next closely enough that downstream structures can respond appropriately without constant reinterpretation. Timing remains usable. Sequencing remains coherent. Recovery remains sufficient. The system still behaves as though it knows what phase it is in.

Unstable variation is different because it introduces uncertainty into the pattern itself. The rhythm may become too uneven, too fragmented, too delayed, too clustered, or too inconsistent in amplitude or timing to remain functionally coherent. The issue is not merely that something has changed. It is that the pattern no longer preserves the same degree of internal order. A system that once oscillated within a recognizable range may begin to show drift, missed transitions, uneven recovery, or exaggerated swings that reduce predictability. This is not just "more variability." It is a different kind of variability—one that reflects reduced coordination rather than preserved adaptability.

That distinction is especially important because variability is often interpreted too positively in broad physiological discourse, as though any departure from rigid regularity must represent flexibility or responsiveness. Sometimes it does. But variability only reflects adaptability when it remains integrated into a larger stable pattern. If the underlying signaling conditions are degrading, variability may instead represent noise, weakened control, or reduced capacity to preserve timing across repeated cycles. Electrolyte instability can contribute to this by altering the baseline conditions on which oscillatory behavior depends. If gradients narrow, if excitability drifts, if recovery becomes less complete, or if signaling becomes less consistent, rhythmic behavior may begin to lose its structure even before overt failure occurs.

Irregularity is often one of the clearest visible signs that a deeper regulatory structure is becoming unstable. A rhythm becomes irregular when the system can no longer preserve the temporal coherence it once maintained. The irregularity itself may appear at the level of timing, amplitude, interval, recovery, or sequencing, but in each case it points beyond the surface event. It suggests that the system is no longer returning to the same baseline conditions with the same reliability. The visible pattern becomes less stable because the invisible conditions beneath it have become less repeatable.

This is especially revealing in systems where rhythm is central to coordination. Any tissue or process that depends on cyclical signaling, alternating phases, or repeated activation-release sequences becomes vulnerable when rhythm loses predictability. The immediate concern is not simply that one event has become different. It is that the broader system can no longer organize around a stable pattern. Downstream structures rely on timing. Coordination depends on sequence. If oscillatory behavior becomes less coherent, then communication, recovery, and responsiveness all become harder to preserve. The cost of rhythm instability is therefore not only local irregularity. It is broader loss of temporal coordination.

A stable rhythm in one domain often supports stability in another because physiological systems are temporally linked. Signaling patterns, fluid regulation, contraction cycles, recovery phases, and autonomic behavior all influence one another over time. If one rhythmic domain becomes unstable, the effect may not remain confined to that domain. It may alter the timing or predictability of related systems, increasing the burden on the organism to maintain coherence across a more fragmented temporal landscape. This is one reason electrolyte-related instability can have effects that appear distributed rather than neatly localized. Electrolytes support conditions that are used repeatedly across systems, and repeated use means temporal patterning matters.

This also clarifies why instability should not be reduced to discrete events. A single episode of dysfunction may attract attention because it is visible, but often the more meaningful question is whether it reflects a breakdown in the pattern that preceded it. Systems rarely become unstable all at once without prior changes in rhythm, timing, or repeatability. Irregularity often appears before collapse because rhythm is sensitive to underlying shifts in coordination. A pattern can degrade long before a single event becomes severe enough to stand out in isolation. This makes pattern-based interpretation more informative than event-based interpretation in many cases. The rhythm often tells the truth earlier than the event.

To interpret physiology through pattern rather than event is therefore to ask different questions. Instead of asking only whether something happened, the more useful question becomes how it is happening across time. Is the sequence preserved? Is the interval coherent? Is the oscillation bounded? Is the recovery phase sufficient? Does the pattern remain recognizable under ordinary load, or is it becoming more erratic, more fragmented, or more difficult to predict? These are rhythm questions, and they are often closer to the underlying mechanism than a simple count of visible events. Electrolytes matter here because they help determine whether the system can keep producing recognizable patterns rather than merely occasional outputs.

That recognizable quality is crucial because functional integrity depends on repeatable interpretability. A rhythm does not need to be rigid to be useful, but it must remain coherent enough that the rest of the system can respond to it appropriately. If the oscillation becomes too unstable, then each new cycle carries more uncertainty. Recovery may no longer match activation. One phase may bleed into the next. Timing may become too variable to coordinate around. These changes do not always announce themselves dramatically at first. Often they emerge as a gradual erosion of predictability. But predictability is not cosmetic. It is one of the main ways physiological systems remain governable across time.

Seen clearly, rhythm stability is not a specialized feature of only a few visible systems. It is a general expression of regulated physiology. Stable systems oscillate. They rise and fall, activate and release, respond and recover, but they do so in ways that remain bounded and intelligible. Electrolyte balance supports this by preserving the gradients, signaling consistency, and coordination that make those oscillations possible without excessive drift or fragmentation. When those supporting conditions remain intact, variability can remain adaptive because it occurs within pattern. When they degrade, the pattern itself begins to lose coherence, and what appears on the surface as irregularity is often the temporal expression of deeper instability in the system beneath it.

Loss

Electrolytes are continuously lost under stress conditions. That is the governing mechanism. Electrolyte loss is not best understood as a rare, dramatic, or exceptional event that occurs only under obvious extremes. It is better understood as an ongoing physiological process shaped by ordinary regulation, environmental exposure, activity, altered input states, and the continuous demands placed on fluid distribution and signaling systems. Electrolytes are always moving, always being retained or excreted in different proportions, and always participating in conditions that can shift under load. Loss therefore belongs to the ordinary logic of physiology rather than to the category of crisis alone. What changes under stress is not whether loss exists, but how rapidly it occurs, through which pathways it becomes more consequential, and how well the system can absorb it without drifting into instability.

This distinction matters because loss is often interpreted too discretely. In common understanding, something is lost when it leaves the body visibly or abruptly, or when a large and obvious depletion event can be named. But physiological systems rarely operate in such clean episodes. Electrolyte loss often unfolds as a layered process: some of it visible, some of it measurable only through changing concentration or distribution, some of it acute, some of it cumulative. Sweat leaves the surface. Urine leaves the body. Fasting alters handling and excretion patterns. But these are not simply separate exits through which electrolytes disappear in a straightforward arithmetic way. They are regulatory pathways through which the system is continuously rebalancing itself under changing conditions. Loss must therefore be understood not just as departure, but as the shifting of a regulatory landscape.

The most visible pathways of loss are sweat and urine, and each requires careful interpretation because each reflects more than simple subtraction. Sweat is often imagined as water leaving the body, but it is also a route through which electrolytes are carried out of the extracellular environment. This matters because sweat loss is not compositionally neutral. The fluid leaving the surface is not identical in all contexts, and its significance depends not only on how much volume is lost but on what that loss means for extracellular concentration, fluid partitioning, and the system's effort to preserve gradients under changing heat and activity conditions. Sweat therefore represents not just fluid departure, but a shift in the extracellular environment that may become progressively more relevant as exposure continues.

Urinary loss is less visible but often more structurally significant because it is not merely a passive leak. It is one of the primary means by which the body regulates fluid and electrolyte balance over time. Urine does not simply carry away excess in a neutral sense. It reflects ongoing decisions made through renal handling, hormonal signaling, fluid status, and broader system priorities. Sodium, potassium, and other electrolytes are not excreted randomly. Their excretion changes according to context, and those changes alter concentration and distribution even when they do not produce immediate visible symptoms. Urinary loss is therefore not best understood as an endpoint. It is part of the continuous regulatory conversation through which the system tries to preserve stability while adapting to load, intake, stress, and changing internal conditions.

Fasting-related shifts add another important dimension because they show that electrolyte loss is not always a direct consequence of external output alone. When intake patterns change, hormonal conditions change, renal handling changes, and fluid distribution often changes with them.

During fasting, electrolyte dynamics are not defined only by the absence of incoming material. They are shaped by altered excretion patterns, shifting extracellular conditions, and the system's changing regulatory priorities. This means that loss can accelerate or take on different significance even when the most obvious external conditions appear quiet. A person may not be visibly sweating heavily or engaging in obvious exertion, yet the underlying handling of electrolytes may still be changing in ways that matter over time.

This is why loss must be understood through context rather than through fixed assumptions. Environmental conditions alter the pace and significance of electrolyte loss because they change how much fluid must be moved, how much heat must be dissipated, how much pressure the system is under to preserve volume and signaling stability, and how much regulatory compensation is required to maintain function. Heat is the clearest example because it amplifies sweat-mediated loss and changes the burden placed on extracellular volume preservation. But environmental load is not limited to heat alone. Dryness, altitude, illness, and any condition that alters fluid handling or excretory demand can change the pattern of electrolyte loss even when the surface event appears ordinary.

Activity introduces another layer because it increases both turnover and demand. A system under physical load does not simply use more energy. It also shifts fluid, generates heat, alters sweat output, changes renal handling, and places repeated demand on signaling, contraction, and recovery cycles that all depend on electrolyte-supported stability. This means that activity does not only increase the need for electrolytes in an abstract sense. It often changes the pattern of how they are lost, redistributed, and regulated. The consequence is that loss under activity is not reducible to the visible quantity of sweat or to the immediate aftermath of exertion. It includes the cumulative strain placed on the regulatory systems trying to preserve concentration and gradients while the environment inside and outside the body remains in motion.

This is where acute and gradual loss must be separated conceptually. Acute loss is easier to recognize because it is associated with clear and often compressed shifts: heavy sweating, rapid fluid change, obvious excretion, abrupt fasting-related adjustments, or a sudden environmental demand that changes output quickly. Acute loss matters because it can alter concentration and distribution over a short period, sometimes faster than the system can fully compensate. But acute loss is only one form of instability. It is often overemphasized precisely because it is visible. What receives less attention, but may be equally important over time, is gradual cumulative loss.

Gradual loss is more difficult to interpret because it rarely presents as a single defining event. Instead, it accumulates through repeated small shifts that individually appear manageable but collectively alter the system's operating conditions. Repeated sweating without obvious collapse, modest urinary shifts across days, altered fasting-related handling, mild environmental exposure, and repeated low-grade regulatory burden can slowly change extracellular volume, intracellular concentration, and signaling conditions without producing a dramatic moment of recognition. This is one of the main reasons depletion is often missed. The system can compensate for a long time. Surface function may remain largely intact. But compensation is not the same as preserved margin.

This is where invisible depletion becomes especially important. Invisible depletion refers not to something mysterious, but to a state in which loss has progressed enough to alter concentration, distribution, or functional reserve without yet producing a dramatic or unmistakable event. The absence of obvious symptoms does not mean the absence of meaningful change. A system can lose margin quietly. Extracellular concentration can narrow. Intracellular conditions can become less stable. Signaling thresholds can become slightly less reliable. Recovery can become slightly less complete. None of these changes needs to be catastrophic to matter. In many cases, the earliest signs of electrolyte-related instability are not dramatic failures, but subtle reductions in the system's ability to absorb further load without drift.

This is one reason loss should not be interpreted as simple subtraction. When electrolytes leave through sweat or urine, or when fasting alters excretion patterns, the resulting instability is not explained by the missing quantity alone. What matters is how that loss changes concentration relative to fluid, how it alters compartmental distribution, and how it increases the burden on regulatory systems attempting to preserve gradients under changing conditions. A small loss under one set of circumstances may be functionally negligible. The same loss under a different fluid status, environmental exposure, or regulatory constraint may be more consequential. This is because loss is not meaningful in isolation. It becomes meaningful through its relationship to the system that must absorb it.

That system-level relationship is what makes cumulative loss especially easy to overlook. Because the body is adaptive, it often preserves visible function while narrowing internal margin. Hormonal regulation shifts. Renal handling changes. Fluid distribution adjusts. Extracellular volume may be defended temporarily at the expense of other domains. Intracellular stability may be partly preserved while extracellular concentration becomes more strained. The result is that loss may be ongoing without appearing dramatic, and the absence of immediate dysfunction may falsely suggest that nothing significant is changing. But stability maintained through increasing compensation is not the same as stability maintained with ease. The distinction matters because a system already working harder to preserve its baseline is less tolerant of further load.

This is also why environmental and activity-driven variation must be taken seriously as part of ordinary physiology rather than treated as edge cases. Electrolyte loss does not happen in a vacuum. It changes according to climate, exertion, fasting state, illness, and the broader regulatory context in which the body is operating. Two days may involve the same nominal intake and yet produce different internal consequences if one includes heat, stress, altered sleep, repeated sweat exposure, or a shift in fluid handling. The system does not lose electrolytes at a fixed rate independent of context. It loses them in relation to conditions. This means that understanding loss requires attention not only to what exits, but to what conditions were present while it exited.

This context sensitivity also explains why loss is often misread when interpreted only through overt discomfort or crisis. A person may not feel acutely unstable and yet may still be moving through a progressive reduction in reserve. The system may continue functioning adequately while becoming less buffered, less electrically stable, or less capable of preserving concentration under additional demand. Because this process can unfold quietly, it is often only recognized retrospectively, after a larger disruption reveals that the margin had already narrowed. But the disruption itself is not always the beginning of the problem. Often it is the first visible consequence of an ongoing process that had been unfolding across many smaller losses.

Loss is therefore not best understood as a discrete interruption of normal physiology. It is part of normal physiology, made more consequential under stress, environmental load, altered input states, and repeated demand. Electrolytes are always in motion, always subject to excretion, redistribution, and concentration change. Stability depends on how well the system can regulate this motion over time without allowing gradual or acute loss to erode the gradients and compartmental conditions on which signaling and fluid balance depend. The critical point is not simply that electrolytes can be lost. It is that they are being lost continuously, in patterns shaped by context, and that these patterns alter the system long before loss becomes dramatic enough to be mistaken for a single event.

Imbalance

Instability emerges from mismatch between input and demand. Electrolyte imbalance is often misunderstood because it is commonly framed as a problem of too little or too much, as though physiology were governed primarily by isolated quantities crossing simple thresholds. But imbalance is more accurately understood as misalignment within a dynamic system. Input, retention, loss, distribution, concentration, and system demand are all moving at once, and stability depends on whether those movements remain coordinated. An electrolyte state can become unstable even when total quantities appear materially adequate if the system is retaining poorly, losing disproportionately, distributing unevenly, or operating under demand conditions that exceed what current availability can support. The issue is not merely amount. The issue is whether the system's requirements and its regulatory conditions still match.

This is the necessary starting point because input alone is too often treated as the main explanatory variable. Intake is visible, countable, and easy to narrate. It lends itself to a simple model in which what enters the system determines what is available to the system. But physiological availability is not defined at the moment of entry. What matters is not only what comes in, but what is retained, where it is held, what is lost, what is diluted, what is shifted between compartments, and how much demand is being placed on the system at the same time. Intake is therefore only one term in a larger equation. It is the beginning of a process, not the answer to it.

Retention is what gives input physiological relevance over time. An electrolyte entering the system becomes meaningful only if it remains available within the compartments and concentration relationships that preserve function. Retention is not merely storage. It is the regulated preservation of usable availability. Sodium retained in a way that supports extracellular volume and pressure is not equivalent to sodium simply present somewhere in total load. Potassium retained in a way that preserves intracellular stability is not equivalent to potassium existing materially in the organism without regard to compartmental placement. Magnesium retained in a functionally meaningful state is not equivalent to magnesium counted in gross terms without reference to its modulatory role. Retention is therefore not a secondary detail. It is one of the central determinants of whether input becomes stability or remains physiologically irrelevant.

Loss, by contrast, is not simply the opposite of input. It is an active and continuous part of the same dynamic system. Electrolytes are lost through sweat, urine, altered fasting-related handling, and broader regulatory shifts, but the significance of that loss depends on timing, volume, concentration, and context. A loss event is not meaningful only because something exits. It becomes meaningful because it changes what remains available relative to demand. This is why imbalance cannot be understood by looking at intake or loss in isolation. A given loss pattern may be tolerable under one set of retention and demand conditions, and destabilizing under another. The system is not governed by single variables acting alone. It is governed by the relationship among them.

That relationship is dynamic rather than static. At any given time, the system is balancing what is coming in, what is being held, what is being lost, what is being redistributed, and what current functional load requires. This means balance is not a fixed inventory state. It is a continuously negotiated alignment. When that alignment is preserved, stability can be maintained even under shifting conditions. When it is not, imbalance begins to emerge, often before any single variable looks dramatically abnormal in isolation. This is one reason electrolyte instability is so often misread. The system can become misaligned gradually while individual inputs or outputs still appear superficially plausible when viewed one at a time.

Ratio disruption is one of the clearest ways this misalignment becomes functionally important. Electrolytes do not operate independently. Their meaning depends in large part on how they relate to one another across compartments and within signaling environments. Sodium and potassium form a central transmembrane relationship that supports electrical asymmetry, excitability, and directional signaling. Magnesium influences how activation is modulated and how signaling remains governable. These ions therefore do not simply contribute separate quantities to the same general pool. They participate in a relational structure. When one shifts disproportionately relative to another, the result is not merely a change in amount. It is a change in system behavior.

This is why imbalance should not be reduced to isolated low or high states. A sodium concentration may look acceptable in abstraction while the ratio between sodium and surrounding fluid has become less stable. Potassium may be materially present while its relationship to sodium across the membrane has narrowed enough to alter excitability. Magnesium may not be absent in a gross sense while its modulatory role has become functionally constrained relative to system activation. In each case, the issue is not simply what exists. It is what exists in relation to something else. Ratios matter because signaling, fluid partitioning, threshold behavior, and recovery all depend on relative structure, not just on isolated totals.

That relative structure becomes especially important under changing demand. Demand is often the least appreciated variable in discussions of imbalance because it is harder to quantify than intake or excretion, yet it is often what determines whether a given electrolyte state remains stable or becomes insufficient. A system under low demand may preserve function despite relatively narrow margin. A system under high signaling load, environmental stress, repeated sweating, fasting-related shifts, illness, or sustained activation may require more from the same underlying availability. This does not mean demand changes the chemistry in a mystical way. It means that the same pool of retained and distributed electrolytes must now support more cycles, more regulation, more buffering, or more fluid and signaling stability than before.

Imbalance emerges when electrolyte availability no longer matches what the system is being asked to do. This mismatch can occur even if total quantities have not obviously collapsed. A person may still possess substantial sodium load while extracellular volume becomes harder to preserve under heat or repeated fluid loss. Potassium may still be present while intracellular signaling becomes less stable under ongoing excitatory demand or distributional drift. Magnesium may still exist materially while modulation becomes less effective in a system under cumulative activation pressure. The key issue is not simply whether the electrolyte is present. It is whether it remains available in the right form, place, and relationship to meet current system demand.

This is where effective availability must be separated from raw presence. Effective availability refers to what the system can actually use within the compartments, ratios, and signaling contexts that matter. Raw presence refers to what exists in total. These are not equivalent. An electrolyte can be present but poorly available because it is diluted, misplaced, disproportionately lost from one compartment, or retained in a way that does not support the functions currently under strain. This distinction is crucial because it explains why imbalance often develops in ways that are not immediately obvious. The material may still be there, but the system may no longer be able to deploy it where and when coordination requires it.

Partial balance is one of the main reasons this becomes difficult to detect. A system can preserve relative stability in one compartment or function while broader imbalance is already developing elsewhere. Extracellular volume may be defended for a time while intracellular signaling margin narrows. Concentration may be held within one domain through compensation while ratios become more distorted in another. Renal handling may preserve one variable at the expense of increasing regulatory strain elsewhere. This means a person or system can appear partially stable while actually becoming more globally misaligned. Balance in one visible layer is not proof of system-wide coherence.

This is one reason isolated measures can be misleading when interpreted without context. A single concentration, a single intake estimate, or a single visible output does not necessarily capture whether the broader system remains aligned. Electrolyte balance is not a single-point property. It is a network property emerging from coordinated relationships across time. The system may temporarily preserve one measurement through compensation even while deeper instability accumulates in gradients, ratios, or compartmental availability. This is not because physiology is opaque for its own sake. It is because stability is relational, and relations are harder to reduce to one isolated indicator without losing the mechanism that actually matters.

This also explains why imbalance is often subtle rather than immediately dramatic. The body does not usually move from perfect alignment to overt failure in one step. More often, it moves through phases of narrowing margin. Demand begins to outpace retention slightly. Loss begins to exceed replacement slightly. Ratios begin to shift. Distribution becomes less ideal. Compensation increases. The visible surface may still appear largely intact because the system is adaptive and can preserve ordinary function for a time. But adaptation under strain is not the same as preserved stability. A system can continue operating while becoming less buffered, less precise, and less tolerant of further perturbation.

This narrowing of margin is often where the most meaningful instability lives. A person may not yet show overt collapse, and the absence of dramatic symptoms may falsely imply that balance has been preserved. But if retention is weakening, loss is increasing, ratios are drifting, or demand has quietly risen beyond what the system can comfortably support, imbalance is already present in structural terms. It may not yet be visible as a severe event, but it is already altering how the system behaves under load. This is why waiting for obvious dysfunction to define imbalance is conceptually too late. By the time dysfunction is unmistakable, the misalignment has often been present for some time.

The more accurate interpretation, then, is to understand imbalance as a condition of miscoordinated relationships. Input no longer matches loss. Retention no longer supports effective availability. Ratios no longer preserve signaling coherence. Demand no longer fits the system's current electrolyte state. One compartment may remain temporarily defended while another narrows. The visible whole may still appear stable while internal coordination becomes more strained. This is not deficiency in the narrow sense, nor is it excess in the simple sense. It is a system whose variables are no longer aligned well enough to preserve functional integrity with the same ease as before.

Seen clearly, electrolyte imbalance is not a discrete category separate from normal physiology. It is what emerges when the ordinary relationships that sustain stability begin to drift out of coordination. Because those relationships are dynamic, imbalance can develop gradually, quietly, and relationally rather than all at once or in a single obvious direction. This is why total quantity alone is too blunt a measure, why intake alone is too incomplete a story, and why isolated signs often fail to capture the full structure of the problem. Stability depends on alignment across input, retention, loss, distribution, ratios, and demand. When that alignment weakens, instability begins—not always loudly, but often precisely where the system has become least able to hide the mismatch any longer.

Compensation

The system attempts to maintain stability under disruption. That is the governing mechanism. Physiological systems do not move directly from stability to failure when conditions shift. They respond. They adjust. They redistribute. They regulate. Compensation is not an exception to normal behavior but a defining feature of it. When electrolyte conditions begin to drift, when loss increases, when distribution becomes less ideal, or when demand rises beyond prior levels, the system does not remain passive. It actively reorganizes internal relationships in an effort to preserve functional integrity. Stability, in many cases, is not the absence of disturbance. It is the visible result of ongoing compensatory activity.

This activity operates across multiple layers simultaneously. Fluid distribution shifts between compartments. Renal handling adjusts retention and excretion patterns. Hormonal signals alter how sodium is preserved or released, how potassium is balanced, and how volume is maintained. Cellular transport mechanisms continue to work to preserve gradients even as external conditions change. None of these processes are incidental. They are coordinated responses designed to maintain a usable internal environment despite deviation from ideal conditions. Compensation, in this sense, is not an emergency response. It is the system's default mode of preserving order under load.

To understand compensation properly, it must be separated from the idea of correction. Correction implies that the underlying imbalance has been resolved, that the system has returned to a baseline state without additional strain. Compensation, by contrast, often involves redistribution rather than resolution. Electrolytes may be shifted from one compartment to another. Retention may increase to offset loss. Excretion may decrease in an attempt to preserve availability. Fluid may be moved to support pressure and volume at the expense of other domains. These adjustments can restore outward function temporarily, but they do not necessarily eliminate the underlying mismatch that prompted them.

This distinction between redistribution and correction is central because it explains why apparent stability can coexist with ongoing strain. A system may maintain blood pressure, preserve signaling coherence, or continue producing coordinated output even as the internal conditions supporting that output become more difficult to sustain. The visible function appears stable, but the stability is being actively maintained through adjustments that carry cost. Compensation is therefore not free. It requires energy, regulatory effort, and trade-offs across systems. What appears as continuity on the surface may reflect increasing internal constraint.

Short-term stability often depends on this kind of compensatory adjustment. When electrolyte loss increases, retention may rise to preserve extracellular volume. When gradients begin to narrow, transport activity may increase to maintain electrical asymmetry. When signaling conditions become less ideal, modulation may intensify to prevent excessive activation or drift. These responses can preserve function effectively for a period of time. The system remains operational. Cycles continue. Coordination is maintained. From an external perspective, little appears to have changed. This is why compensation can be so effective. It allows the organism to continue functioning under conditions that are no longer optimal.

However, this effectiveness is also what makes compensation difficult to detect. Because function is preserved, it is easy to interpret that preservation as evidence that nothing is wrong. The system appears stable, so it is assumed to be stable in a deeper sense. But the stability being observed may not reflect a return to equilibrium. It may reflect a system working harder to maintain its baseline in the face of ongoing disruption. Compensation masks imbalance by preserving output while the underlying conditions remain altered. The masking is not deceptive in intent. It is a consequence of the system's capacity to adapt.

Masking becomes more pronounced when compensation is prolonged. Over time, the adjustments required to maintain stability can become more extensive. Redistribution may become more persistent. Retention mechanisms may remain elevated. Loss pathways may be suppressed or altered. Signaling conditions may operate within narrower margins. These changes allow the system to continue functioning, but they also mean that the distance between apparent stability and true alignment increases. The system is no longer operating from a position of ease. It is operating from a position of managed imbalance.

This managed imbalance introduces strain. Strain is not always visible immediately, but it accumulates as the system continues to compensate. Each adjustment places demand on regulatory pathways. Each redistribution changes how resources are allocated. Each attempt to preserve one domain may constrain another. The system remains coordinated, but the coordination becomes more conditional. It depends on continued compensatory effort rather than on naturally aligned conditions. Over time, this can narrow the range within which the system can operate without further adjustment.

This narrowing of range is one of the clearest signs that compensation is approaching its limits. A system with wide margin can absorb variation without significant disruption. A system operating under prolonged compensation has less flexibility. Small additional stresses may produce disproportionate effects because the system is already near the edge of its compensatory capacity. The visible output may still appear stable, but the underlying tolerance has decreased. The system becomes more sensitive to further loss, further demand, or further shifts in distribution because it has less room to adjust.

This is why compensation should not be interpreted as a permanent solution. It is a process that buys time, preserves function, and delays overt instability, but it does not eliminate the underlying conditions that required it. If those conditions persist, compensation must continue, and if compensation must continue, strain accumulates. The system does not fail immediately because it adapts, but adaptation under sustained load has limits. These limits are not defined by a single threshold. They are defined by the point at which the system can no longer maintain coordination without sacrificing functional integrity.

When those limits are approached, the masking effect of compensation begins to weaken. Instability that was previously contained may become more visible. Rhythms may become less predictable. Signaling may become less consistent. Recovery may become less complete. Fluid distribution may become more difficult to regulate. The system is not suddenly failing without warning. It is transitioning from a state in which compensation was sufficient to preserve outward stability to a state in which compensation can no longer fully contain the underlying imbalance. What becomes visible at this stage is not a new problem. It is the exposure of a process that had been ongoing beneath the surface.

This transition underscores the importance of understanding compensation as a dynamic and limited process. It is not a binary state in which the system is either compensating or not. It is a continuum. Early compensation may involve minimal adjustment with little strain. Intermediate compensation may involve more extensive redistribution and increased regulatory effort. Late compensation may involve significant constraint, reduced margin, and increased vulnerability to further disruption. The system can move along this continuum gradually, often without a clear dividing line between stages. What changes is not the presence of compensation, but its intensity and its cost.

One of the common errors in interpreting physiological behavior is to assume that as long as function is maintained, stability has been preserved in a complete sense. This assumption overlooks the possibility that function may be maintained through compensation rather than through alignment. It confuses outcome with mechanism. A system can produce stable output through active adjustment while its underlying conditions remain misaligned. Recognizing this distinction requires attention not only to what the system is doing, but to how it is doing it. Stability achieved through minimal strain is different from stability achieved through continuous compensatory effort.

This distinction also clarifies why compensation can obscure the progression of imbalance. Because the system continues to function, the changes occurring beneath the surface may go unrecognized. Redistribution may alter concentration relationships. Retention may shift electrolyte availability across compartments. Loss may continue at a rate that gradually exceeds replacement. Demand may remain elevated. These processes can coexist with preserved function, creating the impression that no significant change is occurring. In reality, the system may be moving steadily toward a state in which compensation is no longer sufficient.

Understanding compensation as an active, limited, and strain-bearing process provides a more accurate framework for interpreting stability and disruption. It highlights that the absence of overt failure does not necessarily indicate the absence of imbalance. It clarifies that redistribution is not equivalent to correction. It shows that stability can be maintained temporarily under conditions that are not sustainable indefinitely. And it reveals that the transition from apparent stability to visible instability is often the culmination of a prolonged period of compensatory activity rather than a sudden and isolated event.

Stability Envelope

Electrolytes operate within a narrow stability band. That is the governing mechanism. Physiological systems do not function across unlimited ranges, nor do they require exact immobility to remain stable. Instead, they operate within defined boundaries—upper and lower thresholds—inside which variation is not only tolerated but necessary. This bounded region can be understood as a stability envelope, a constrained range within which gradients, signaling, fluid distribution, and coordination remain functional. Outside of this range, the same variables that support stability begin to undermine it. Electrolyte-driven systems therefore depend not on fixed values, but on maintaining oscillation within limits that preserve functional integrity.

These limits are not arbitrary. Upper and lower thresholds reflect the conditions under which the system can still interpret signals, maintain gradients, and coordinate across compartments and time. Below the lower threshold, concentration, pressure, or electrical potential may fall to levels that no longer support effective signaling or distribution. Above the upper threshold, excess may distort gradients, alter osmotic relationships, or disrupt the same signaling systems through overamplification or misalignment. In both directions, the issue is not simply too little or too much in a descriptive sense. It is that the system's operating conditions have moved beyond the range in which coordinated behavior remains possible.

This framing clarifies that stability is not a single point of equilibrium. It is a range. Within that range, oscillation occurs continuously. Electrolyte concentrations shift, fluid moves between compartments, signaling rises and falls, and gradients are used and restored. These changes are not signs of instability. They are the mechanism of normal function. A stable system is one in which these oscillations occur within the stability envelope, where each fluctuation remains bounded enough that the next phase can proceed without cumulative distortion. The system is not static. It is dynamically contained.

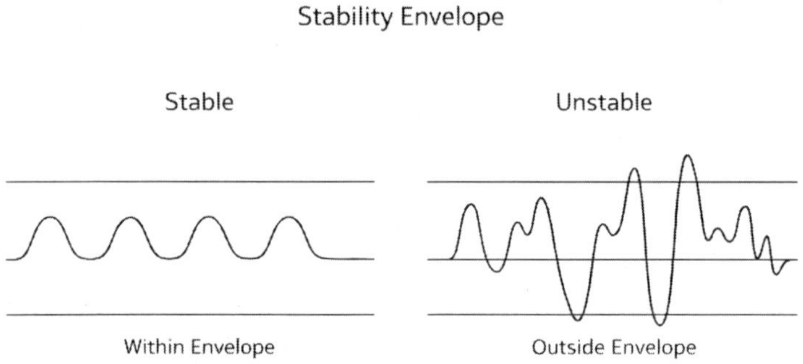

[Figure 5: Stability envelope showing oscillations maintained within thresholds versus excursions beyond the boundaries that produce unstable behavior.]

This bounded oscillation is what allows systems to remain responsive while preserving coherence. A signal can rise, propagate, and resolve because the underlying conditions remain within limits that support repeatable transitions. Fluid can shift without collapsing compartmental balance because gradients remain within usable ranges. Electrical activity can fluctuate without becoming erratic because thresholds are not crossed in ways that alter excitability unpredictably. The stability envelope is therefore not a restriction imposed on the system. It is the condition that allows the system to remain both flexible and controlled.

Instability begins when movement exceeds these bounds. When electrolyte concentrations, fluid distribution, or electrical conditions cross upper or lower thresholds, the system is no longer oscillating within a contained range. It is operating outside the envelope that supports coordinated behavior. This does not always result in immediate failure. Systems may continue to function in a compromised state for a period of time. But once outside the envelope, signaling becomes less reliable, gradients become harder to maintain, and coordination begins to degrade. The issue is not that variation has occurred. The issue is that variation has exceeded the limits within which variation remains useful.

This is why thresholds should not be understood as arbitrary lines imposed on otherwise continuous systems. They are functional boundaries. Crossing them changes the behavior of the system. A gradient that narrows too far may no longer support directional signaling. A concentration that rises too high may alter osmotic relationships in ways that disrupt fluid balance. An electrical condition that shifts beyond a certain point may change threshold behavior and excitability across tissues. These changes are qualitative, not merely quantitative. The system behaves differently once thresholds are crossed, even if the numerical difference appears incremental.

Within the stability envelope, however, oscillation is not only tolerated but required. The system must move. Electrolytes must shift between compartments. Signaling must rise and fall. Pressure must adjust. Recovery must follow activation. A system that did not oscillate would not be alive in any meaningful sense. Stability therefore cannot be defined as stillness. It must be defined as movement that remains contained. This containment allows each oscillation to complete without pushing the system into a new and less stable operating condition.

This is where the relationship between magnitude and frequency becomes central. Magnitude refers to how far a variable moves within or beyond the stability envelope. Frequency refers to how often these movements occur. Both dimensions matter. A large movement that occurs rarely may be tolerable if the system has sufficient capacity to recover fully between events. A smaller movement that occurs repeatedly may become destabilizing if recovery is incomplete and oscillations begin to accumulate rather than resolve. The stability envelope is therefore not defined only by how large changes are, but also by how often they occur and whether the system can restore baseline conditions between them.

Focusing on magnitude alone leads to an incomplete understanding of stability. A system may remain within acceptable bounds for individual events, yet still become unstable if those events occur with a frequency that exceeds the system's capacity to recover. Conversely, a system may tolerate occasional excursions near the edges of the envelope if they are infrequent and followed by sufficient restoration. Stability depends on the interaction between these dimensions. Oscillations must remain both bounded in magnitude and appropriately spaced in time relative to the system's ability to maintain gradients and restore conditions.

This interaction also explains why wider ranges are not inherently beneficial. It might appear that a system capable of tolerating larger swings would be more robust. But wider oscillations often mean that the system is operating closer to its thresholds more frequently, increasing the likelihood of crossing them under additional load. A tightly regulated envelope allows for controlled variation without approaching the limits of functional integrity. A widened envelope may reflect reduced control rather than increased flexibility, especially if the system begins to operate near or beyond thresholds with greater regularity.

The relationship between oscillation and thresholds also reveals why instability can emerge gradually rather than abruptly. A system may begin to show slightly increased amplitude in its oscillations or slightly increased frequency of fluctuation without immediately crossing thresholds. Over time, however, these changes can bring the system closer to its limits more often, reducing margin and increasing vulnerability to disruption. Eventually, what were previously contained oscillations begin to breach the envelope, and the system transitions from stable variability to unstable behavior. The progression is often continuous, even if the visible effects appear suddenly.

Matching system demand to the stability envelope is therefore essential for preserving coordination. Demand determines how much movement is required within the system. Increased activity, environmental stress, altered intake patterns, or cumulative load all increase the magnitude and frequency of oscillations. For stability to be maintained, the envelope must be sufficient to contain these oscillations, and the system must be able to regulate them effectively. If demand increases without a corresponding capacity to maintain gradients and control oscillation, the system is more likely to exceed its thresholds.

This does not mean that the envelope itself is fixed in all contexts. It can shift within limits depending on adaptation, regulation, and environmental conditions. But it remains bounded. There are still upper and lower thresholds beyond which coordination degrades. The key variable is whether the system can keep its oscillations within those bounds under current conditions. When demand is aligned with the system's capacity, oscillations remain contained, and stability is preserved. When demand exceeds that capacity, oscillations begin to approach or cross thresholds more frequently, and instability emerges.

Understanding stability as an envelope rather than a point also clarifies why single measurements can be misleading. A value taken at one moment may fall within acceptable range, yet the system may still be unstable if oscillations are frequently crossing thresholds outside that moment. Conversely, a value slightly outside the typical range at a single time point may not indicate instability if the overall pattern remains bounded and recoverable. Stability is not defined by isolated data points. It is defined by whether the system's behavior across time remains contained within its functional limits.

This pattern-based understanding aligns with the broader structure of electrolyte-driven physiology. Gradients must be maintained within ranges that support signaling and fluid distribution. Electrical conditions must remain within thresholds that preserve excitability and recovery. Oscillations must occur in a way that allows repeated cycles without cumulative distortion. The stability envelope integrates all of these dimensions into a single conceptual framework. It shows that stability is not about eliminating change, but about containing it.

When that containment is preserved, the system can absorb variability, respond to demand, and maintain coordination across time. When it is lost, even small additional changes can push the system further outside its limits, amplifying disruption. The transition from stability to instability is therefore not defined by a single event, but by the system's relationship to its thresholds and its ability to keep oscillation within bounds appropriate to its current load.

Symptom Interpretation

Symptoms reflect system signals, not isolated problems. That is the governing mechanism. Symptoms associated with electrolyte instability are often interpreted too literally and too locally, as though each one were a self-contained defect requiring a direct and singular explanation. But symptoms are better understood as outputs of system behavior. They are not usually the primary event. They are the visible expression of underlying shifts in gradients, distribution, signaling, rhythm, compensation, and coordination. Fatigue, dizziness, cramps, irregularity, weakness, drift in responsiveness, or changes in stability are not inherently self-explanatory phenomena. They are interpretive clues emerging from a larger physiological context. To read them accurately, they must be understood not as isolated problems, but as signals produced by systems under changing conditions.

This distinction matters because symptoms are often granted more conceptual certainty than they deserve. A symptom feels immediate, concrete, and undeniable. It is easy to treat that immediacy as proof that its meaning is equally obvious. But the presence of a symptom does not reveal its mechanism automatically. A cramp is real, but it does not explain itself. Dizziness is real, but it does not name its cause. Fatigue is real, but it does not identify which system-level conditions produced it. The symptom is not the explanation. It is the event requiring explanation. Electrolyte-related instability is frequently misread because the visible output is treated as the whole story rather than as a downstream expression of something broader and less visible.

Common symptoms such as fatigue, dizziness, and cramps are useful examples because they are often interpreted through narrow, one-to-one logic. Fatigue is often reduced to energy deficit or under-recovery in a simplistic sense. Dizziness is often collapsed into a generic notion of dehydration or momentary weakness. Cramps are often treated as a direct sign of one specific electrolyte absence. These interpretations are not always entirely wrong, but they are usually incomplete. Each of these symptoms can emerge from multiple forms of system disruption, and each often reflects not a single broken variable, but a pattern of altered coordination involving fluid distribution, pressure stability, signaling consistency, excitability, or compensatory strain. The visible symptom may be singular. The mechanism rarely is.

Fatigue, for example, can be understood more accurately as a reduction in usable system coherence rather than simply as a lack of energy in the abstract. Electrolyte instability can contribute to this through multiple routes. Fluid distribution may become less supportive of stable pressure and circulation. Signaling may become less consistent. Activation and release cycles may become less efficient. Compensation may increase the regulatory cost of maintaining ordinary function. None of these mechanisms needs to produce catastrophic failure to alter the felt quality of the system's output. Fatigue may therefore represent not one missing input, but a system operating with reduced margin, increased strain, or less stable coordination across repeated cycles.

Dizziness follows a similar logic. It is often experienced as a discrete event, but it is better understood as a symptom of altered system behavior in relation to position, pressure, fluid distribution, or signaling stability. The key point is not that dizziness has no physiological basis. It does. The key point is that its meaning lies in the conditions that make the system less able to maintain coherent function under ordinary shifts in demand.

Electrolyte-related changes in extracellular volume, pressure regulation, or broader coordination can all contribute. The dizziness itself is the visible output of a system failing to stabilize some transition or condition cleanly. It is not simply "caused by dizziness." It is caused by a pattern of instability made visible in that moment.

Cramps are perhaps the clearest example of how one-to-one symptom logic can become misleading. Because cramps involve contraction and discomfort, they are often interpreted as direct evidence of one missing electrolyte, most commonly in a simplistic and isolated sense. But cramps are better understood as disturbances in the control of activation and release cycles. They may reflect altered excitability, incomplete relaxation, signaling noise, increased threshold instability, or a narrowed margin for coordinated contraction-release behavior. Electrolytes are certainly involved, but not necessarily as a simple linear absence. The symptom reflects a system that has become less able to regulate muscular signaling and release cleanly. That is a systems problem, not merely a substance problem.

This is where misinterpretation patterns become especially important. One of the most common errors is the assumption that each symptom maps cleanly onto one cause. This is attractive because it simplifies interpretation. It makes physiology appear legible through direct substitution: symptom equals variable. But the body rarely operates through that kind of single-cause transparency. The same symptom can emerge from different mechanisms, and the same mechanism can produce different symptoms depending on context, compensation, timing, and load. Electrolyte-related symptoms are especially vulnerable to misreading because electrolytes influence multiple domains at once—fluid balance, signaling, excitability, pressure regulation, contraction, recovery, and rhythm. A single symptom may therefore represent the visible intersection of several altered conditions rather than one isolated defect.

Another common misinterpretation is the tendency to read symptoms without temporal context. A symptom appearing at one moment is often treated as a self-contained event rather than as part of a pattern unfolding across time. But physiological meaning often resides in recurrence, timing, clustering, and context rather than in the isolated event itself. A single episode of dizziness, fatigue, or cramping may be difficult to interpret in abstraction. But repeated appearance under specific conditions—after fluid loss, under heat, during fasting-related shifts, with repeated activation, or in states of narrowed recovery—begins to reveal something more informative. The symptom gains meaning not because it has become more dramatic, but because it has become part of a recognizable system pattern.

This is why event-based thinking is too narrow for symptom interpretation. If symptoms are treated only as isolated occurrences, much of their meaning is lost. A better approach is pattern-based analysis. Pattern-based interpretation asks how symptoms behave across time, context, and system demand. Do they emerge after loss, under stress, with positional shifts, after repeated activation, or during periods of altered intake and regulation? Do they cluster with other signs of instability? Do they appear as isolated anomalies or as recurring expressions of the same underlying strain? These questions do not replace mechanism. They help locate it. Symptoms become more interpretable when they are read as part of coordinated system behavior rather than as disconnected interruptions.

Pattern-based interpretation also helps distinguish meaningful signal from noise. Not every fluctuation is important. Living systems vary. Sensations change. Minor irregularities occur. If every transient sensation is treated as highly meaningful, interpretation becomes unstable and overly reactive. But the opposite error is equally common: dismissing repeated or patterned symptoms as meaningless because they are intermittent or inconsistent. The challenge is not to eliminate ambiguity completely. It is to recognize that symptoms exist on a spectrum from noise to signal, and that their position on that spectrum depends on pattern, recurrence, context, and relationship to broader system behavior.

Noise, in this context, refers to fluctuations that do not reliably correspond to a coherent underlying instability pattern. They may be brief, non-recurring, weakly contextualized, or disconnected from broader changes in system behavior. Meaningful signal, by contrast, tends to show structure. It recurs under certain loads, aligns with shifts in fluid or signaling conditions, clusters with related symptoms, or becomes more visible as compensation narrows and system coordination degrades. The distinction is not always obvious at first, which is why symptom interpretation requires patience and pattern recognition rather than reflexive one-step explanations. A meaningful symptom is not necessarily the loudest one. It is the one that remains consistent with a deeper system logic.

This is also why variability should not be dismissed automatically. Intermittent symptoms are often treated as less real or less important because they are not constant. But intermittence is often exactly what should be expected in systems that are still compensating. A compensated system may preserve function under some conditions and reveal instability under others. Symptoms may therefore appear only when the system is asked to do something that exposes its narrowed margin—stand quickly, tolerate heat, sustain repeated contraction, manage fasting-related shifts, or recover from fluid loss. Variability in symptoms does not necessarily indicate randomness. It may indicate conditional instability.

Conditional instability is especially important because it helps explain why symptoms often appear situational rather than continuous. A person may feel largely stable in low-demand conditions and noticeably less stable under environmental load, exertion, altered intake, or prolonged activation. This does not mean the system is healthy at one moment and unhealthy at another in a simplistic binary sense. It means the system is operating near enough to its limits that some conditions remain tolerable while others expose the mismatch. Symptoms, in this case, are not arbitrary. They are context-dependent readouts of a system whose coordination is becoming harder to preserve under specific forms of demand.

This framing also protects against the opposite error: assuming that symptoms are purely subjective noise with little physiological meaning. Because many electrolyte-related symptoms are nonspecific and shared across multiple domains, they are sometimes treated as too vague to matter. But vagueness does not mean irrelevance. It means interpretation must remain structural rather than reductive. A nonspecific symptom may still be highly informative if it appears in the right pattern, under the right conditions, and alongside the right broader signs of system strain. The task is not to force precision where none exists. It is to read symptoms in relation to the system that produced them.

Symptom interpretation, then, is not the process of assigning a label to an isolated sensation. It is the process of reading system behavior through its outputs. Electrolyte-related symptoms become intelligible when they are treated as expressions of gradients under strain, distribution under shift, signaling under instability, and compensation under load. This requires moving beyond direct-correction thinking and into coordinated interpretation. A symptom does not necessarily tell the observer what to do. It tells the observer something about how the system is currently behaving. That difference matters because one approach reacts to the surface event, while the other attempts to understand the structure beneath it.

Seen clearly, symptoms are not noise to be ignored or singular messages to be decoded too literally. They are partial but meaningful expressions of physiological coordination and disruption. They gain clarity when placed in temporal, relational, and systems-based context. Fatigue, dizziness, cramps, and similar outputs are not self-contained problems. They are visible points within a larger pattern of regulation, compensation, and instability. Their value lies not in their immediacy alone, but in what they reveal about whether the system is maintaining coherence, narrowing margin, or beginning to lose the capacity to preserve stability under the conditions being imposed on it.

Thresholds

Systems shift from stable to unstable at defined limits. That is the governing mechanism. Physiological systems do not always deteriorate in a smooth, proportional, and continuously visible way. They often absorb change for a period of time, preserve coordination under strain, and continue functioning within narrowing margins until a critical limit is reached. Once that limit is crossed, behavior can change rapidly. A system that appeared mostly stable may begin to behave very differently with what seems, on the surface, like only a small additional shift. This is threshold behavior. It is not an exception to physiology. It is one of the primary ways living systems organize tolerance, adaptation, and failure.

This matters because linear thinking is intuitively attractive but often misleading. If change is interpreted linearly, then a small additional stressor should produce only a small additional effect, and a large disruption should be required before system behavior changes substantially. But this is not how many electrolyte-dependent systems behave. Signaling, fluid distribution, membrane excitability, rhythm stability, and compensatory regulation often remain relatively coherent across a range of disturbance and then change more abruptly once certain thresholds are exceeded. The apparent suddenness of the transition can create the impression of randomness, but the underlying process is often structured and predictable. The system did not fail without warning. It crossed a limit.

To understand thresholds clearly, it is necessary to distinguish them from gradual change. Gradual change refers to movement within a stable operating range. A system may become somewhat more strained, somewhat less efficient, or somewhat more compensatory without losing functional integrity outright. Threshold crossing is different. It is the point at which incremental change produces a qualitative shift in behavior. Below the threshold, the system can still regulate. Above it, regulation begins to fail, coordination becomes less reliable, or the same amount of perturbation produces disproportionately larger effects. The system has not simply changed in degree. It has changed in mode.

This distinction is especially important in electrolyte physiology because many of the systems involved are highly sensitive to narrow ranges. Gradients do not need to collapse completely to become unstable. Membrane behavior does not need to become entirely disorganized to alter excitability. Fluid distribution does not need to become grossly abnormal to begin affecting pressure, signaling, or tolerance. Electrolyte-related instability often emerges when small shifts push a tightly regulated variable beyond the bounds within which the system can continue compensating cleanly. The visible consequence may look abrupt, but the mechanism is usually the culmination of progressive narrowing rather than the arrival of an entirely new event.

Non-linearity is the key concept here. A non-linear system does not respond proportionately across all ranges. It may tolerate significant variation within one zone and become much more sensitive outside of it. Electrolyte-driven physiology is filled with these non-linear relationships. A certain degree of sodium loss may be well tolerated until extracellular volume and pressure regulation narrow enough that additional loss produces a much larger effect. A certain degree of potassium instability may produce little visible change until membrane excitability shifts enough that signaling becomes less predictable. Magnesium-related modulation may remain sufficient under moderate load until rising activation pressure overwhelms buffering capacity and system behavior becomes more erratic. In each case, the system's response is not simply more of the same. It changes character.

This change in character is what often gets mistaken for randomness or fragility. A person or system may seem relatively stable, and then a comparatively small additional burden—heat, exertion, fasting-related loss, sleep disruption, repeated activation, fluid shift—appears to produce a disproportionate disruption. The temptation is to interpret that disruption as sudden, inexplicable, or disconnected from what came before. But in many cases, the more accurate interpretation is that the system had already moved close to a threshold. The final perturbation was not uniquely powerful. It was simply enough to move the system from one side of a functional boundary to the other.

This is where stability margin becomes one of the most important concepts in the chapter. Stability margin refers to the distance between the system's current operating state and the threshold beyond which behavior begins to change qualitatively. It is the usable buffer between ordinary fluctuation and instability. A wide margin means the system can absorb variation, loss, or load without approaching its limits too quickly. A narrow margin means the system is operating closer to those limits, so even modest additional change can produce disproportionate effects. Stability margin is therefore not a visible substance or a single measurable variable. It is a relational property of how much room the system has left before threshold behavior begins to dominate.

This margin is not fixed. It changes according to context. Environmental load, fluid shifts, cumulative loss, signaling demand, sleep disruption, stress, fasting-related excretion, and repeated compensatory effort can all narrow the available buffer. What may be tolerable under one set of conditions may become destabilizing under another not because the system has become fundamentally different overnight, but because its margin has changed. The same perturbation imposed on a system with wide margin may produce little visible effect. The same perturbation imposed on a system with narrow margin may push it across a threshold. This is one reason threshold behavior is so often misunderstood: the visible trigger is interpreted without regard to the invisible margin that preceded it.

Compensation plays a central role in this narrowing because compensation can preserve outward stability while quietly reducing the system's remaining tolerance. A compensated system may still look functional, but it may be operating closer to its thresholds than surface appearance suggests. Retention may be elevated. Redistribution may be ongoing. Signaling may be preserved through increased regulatory effort. Recovery may still occur, but less completely. These adjustments can keep the system stable enough to continue functioning, but they often do so by using up margin. The system remains intact, but less buffered. It is still coordinated, but more conditionally so.

This is why sudden breakdown is often not truly sudden. It is frequently the visible phase transition that occurs when accumulated strain and narrowing margin finally meet a perturbation large enough—or simply timely enough—to exceed tolerance. The transition may appear abrupt because the visible symptoms or dysfunction become obvious quickly. But the system's movement toward that transition was often gradual and structured. Threshold behavior compresses the visible timeline. It makes the change appear immediate even when the preparation for that change has been unfolding for much longer.

Understanding this helps correct one of the most common interpretive errors in physiology: the assumption that if a system appears mostly stable, it must be far from failure. In reality, a system can appear stable while operating very close to its limits. The difference between preserved function and overt disruption may not be large in quantitative terms when the stability margin is already narrow. This does not mean physiology is fragile in a simplistic sense. It means physiology is bounded. Systems can tolerate only so much deviation before the conditions supporting coordination begin to fail. Thresholds are how those limits become visible.

Tolerance, then, must be understood as finite and dynamic rather than unlimited or static. A system's tolerance for fluid loss, signaling strain, environmental load, or oscillatory disruption is not fixed once and for all. It depends on current distribution, current gradients, current compensatory burden, and current demand. A person may tolerate one condition easily at one time and poorly at another because the underlying stability margin has shifted. This is not inconsistency in the trivial sense. It is the expected behavior of a bounded, adaptive system whose tolerance depends on its current position relative to its thresholds.

This boundedness also clarifies why more is not always better and why wider tolerance is not always the right conceptual goal. If a system appears to tolerate increasingly large swings, that may reflect resilience—but it may also reflect altered regulation, reduced sensitivity to warning signals, or movement toward more unstable oscillation. Stability is not defined by the capacity to absorb unlimited disturbance. It is defined by the capacity to remain coordinated within appropriate bounds. Thresholds are part of what give those bounds meaning. They are not evidence of weakness. They are the structural conditions that make function possible in the first place.

Electrolyte-related systems are especially dependent on these boundaries because they operate through gradients, membrane potentials, compartmental relationships, and signaling conditions that require precision more than excess. Too little destabilizes. Too much destabilizes. Too frequent a shift destabilizes. Too large a swing destabilizes. The point is not that the system cannot change. The point is that change must remain within the limits where gradients still direct, signals still propagate, fluid still distributes coherently, and compensation does not have to become the primary mode of survival. Thresholds are where that coherence begins to give way.

Seen clearly, thresholds do not represent mysterious points of collapse. They represent the limits at which a system can no longer preserve the same form of behavior under current conditions. Below them, regulation still contains the disturbance. Beyond them, the system begins to operate differently. The transition may feel abrupt, but it is often the predictable expression of narrowed stability margin, accumulated load, and non-linear system behavior. Electrolyte imbalance contributes to this process not only by causing obvious depletion or excess, but by shrinking the buffer between ordinary function and instability until the next shift—sometimes small, sometimes ordinary—is enough to move the system across the line where coordination can no longer be maintained in the same way.

Cascading Failure

Local imbalance propagates into system-wide instability. That is the governing mechanism. Electrolyte disruption rarely remains confined to the site or variable where it first becomes visible. Because gradients, fluid distribution, electrical behavior, and signaling thresholds are interdependent, a disturbance in one area alters conditions in others. What begins as a local instability can therefore spread through connected systems, not because the body is fragile in a simplistic sense, but because it is coordinated. Coordination is what allows function to emerge across compartments and tissues. It is also what makes disruption transmissible. When one part of the system becomes unstable, it changes the operating conditions for the parts linked to it.

This means failure is not best understood as a single-point event. A system does not usually move from intact to broken in one step. It more often moves through a process of propagation in which one local disturbance produces several downstream effects, each of which alters the conditions under which other processes are trying to remain stable. A narrowed sodium gradient may alter extracellular volume and pressure behavior. That change may affect signaling reliability, positional tolerance, or perfusion dynamics. Those changes may then increase strain on compensatory pathways, alter excitability, and shift how the system handles subsequent demand. What began as a localized imbalance has now become a broader disturbance in operating conditions.

This is the essential structure of cascading failure. A first disruption changes the background against which other systems are functioning. Those systems, now operating under altered conditions, begin to produce less stable output themselves. Their altered output then becomes a new source of disruption for additional systems. The failure does not spread because one variable somehow "causes everything." It spreads because interconnected systems cannot remain fully isolated from changes in the conditions they share. Electrolytes are central to this process because they participate in precisely those shared conditions: fluid partitioning, electrical asymmetry, membrane behavior, excitability, threshold control, and signal propagation.

A single local disruption therefore has significance beyond its immediate site. Consider a disturbance in extracellular volume regulation. On the surface, this may appear to be a fluid issue. But extracellular volume is not only about quantity of fluid in space. It is also about pressure relationships, transport conditions, tissue perfusion, and the environment in which signaling systems are trying to function. A local disturbance here changes the context for electrical stability, positional regulation, renal response, and compensatory activation. Each of these systems may initially continue functioning, but now from a less stable baseline. The first disruption has already propagated, even if overt breakdown has not yet occurred.

The same logic applies to electrical instability. A shift in membrane behavior or excitability may initially appear localized to one tissue or signaling domain, but the consequences are rarely confined there. Signal propagation becomes less reliable. Threshold behavior becomes less predictable. Cycles of activation and release become harder to coordinate. This can alter muscular behavior, nervous system output, autonomic patterning, and downstream regulatory responses. The original disturbance may have begun in a narrow band of electrical conditions, but the consequences quickly become broader because signaling is one of the primary means through which systems coordinate their timing and response. If signaling becomes less coherent in one domain, other domains begin operating on degraded information.

This is why feedback loops become central to understanding instability. A feedback loop exists when the consequences of a disruption feed back into the conditions that produced the disruption in the first place, either stabilizing it or amplifying it. In electrolyte-related instability, amplification loops are especially important. A disturbance in fluid distribution may alter pressure stability, which increases compensatory activation, which changes renal handling or signaling behavior, which further disturbs distribution. A disturbance in excitability may alter contraction or rhythm, which increases system strain, which narrows gradients further, which increases excitability instability again. These loops are not separate events stacked one after another. They are self-reinforcing interactions that make instability harder to contain once it begins.

Amplification is what transforms a manageable disturbance into a broader instability pattern. Without amplification, a local disruption might remain limited and resolve within the system's ordinary tolerance. With amplification, the consequences of the original disruption begin to strengthen the very conditions that keep it active. This is why some failures seem to accelerate once they begin. The system is no longer dealing only with the original disturbance. It is now dealing with the secondary disturbances created by its own destabilized responses. Each additional disturbance adds another layer of difficulty to the task of preserving coordination.

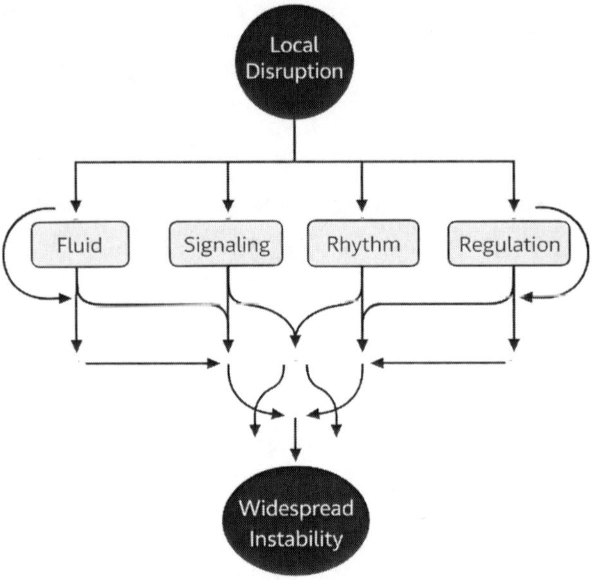

[Figure 6: Local disruption propagating into system-wide instability through branching effects and feedback amplification.]

Compounding failure emerges when several disrupted processes begin interacting simultaneously. At this stage, the system is no longer dealing with one unstable variable spreading outward in a clean sequence. It is dealing with multiple partially unstable domains influencing one another at the same time. Fluid distribution may be less stable. Signaling may be noisier. Compensation may be more active. Recovery may be less complete. Rhythmic behavior may become less predictable. Each of these changes may still be partly compensated, but together they create a system whose internal conditions are becoming harder to coordinate across time. The problem is no longer a single imbalance. It is the interaction between imbalances.

This interaction matters because physiological systems do not fail additively. They fail through combinations. Two mildly destabilized domains may produce far more than twice the disruption of one. This is another form of non-linearity. A moderate disturbance in signaling may be tolerable. A moderate disturbance in fluid distribution may also be tolerable. But when both occur together, their interaction may push the system into a qualitatively different state because each reduces the margin available to absorb the other. Compounding effects therefore reflect not only the number of disrupted processes, but the way those processes interact to narrow overall system tolerance.

As compounding increases, coordination begins to degrade more visibly. Coordination is the system's ability to maintain coherent relationships across compartments, tissues, rhythms, and signaling layers. It is what allows one process to occur in the right place, at the right time, with the right magnitude relative to another. Cascading failure disrupts this by making internal conditions less predictable and less mutually supportive. Signals arrive under degraded gradients. Volume is regulated under altered pressure conditions. Recovery attempts occur under incomplete restoration. One domain may still be trying to stabilize while another is already destabilizing again. The result is not merely weakness or inefficiency. It is fragmentation of system behavior.

This fragmentation often appears, from the outside, as a collection of disconnected symptoms or failures. But the apparent fragmentation is itself a sign of lost coordination. Systems that were previously operating as a coherent whole begin to behave more independently, more inconsistently, or more noisily because the shared conditions that once synchronized them have degraded. A rhythm becomes less regular. A signal becomes less reliable. A fluid shift becomes less well contained. A symptom appears in one domain, then another. These are not separate unrelated problems appearing coincidentally. They are the visible expression of a coordinated system losing the conditions required to remain coordinated.

This is why system breakdown should not be interpreted as a sudden and isolated collapse. What appears as breakdown is often the visible endpoint of a progressive cascade that has been unfolding beneath the surface. The reason it seems abrupt is that coordination can mask instability for a time. Compensation can preserve function. Local disturbances can remain partially contained. But once propagation reaches enough domains and feedback amplification becomes strong enough, the system can no longer preserve the appearance of coherence. Breakdown becomes visible not because the process has only just begun, but because the cascade has reached a scale where masking is no longer sufficient.

At that point, the distinction between local and systemic failure becomes less meaningful because the original disruption has already transformed the conditions under which the whole system operates. A disturbance that began as a local imbalance in one gradient, one compartment, or one signaling domain has altered enough interconnected processes that the system is no longer failing in parts. It is failing as a coordinated whole. This does not mean every subsystem stops functioning at once. It means the shared structure that allowed them to function together has degraded to the point where each domain increasingly destabilizes the others rather than supporting them.

This perspective also helps correct the common tendency to interpret breakdown through the lens of the final visible event. The last symptom, the most obvious dysfunction, or the most dramatic disruption is often treated as the "real" problem. But in cascading failure, the final visible event is usually only the last readable expression of a much longer chain of interactions. The more important mechanism is the propagation that preceded it. What matters most is not just what failed last, but how the earlier local instability was allowed to spread through connected systems until the system's overall capacity for coordinated regulation was no longer sufficient.

Seen clearly, cascading failure is not a mysterious collapse or an overstatement of small disturbances. It is the expected behavior of a tightly coordinated system when local instability is allowed to propagate through shared gradients, signaling conditions, fluid relationships, and feedback loops. The system fails progressively, even when the visible timeline feels compressed. A disturbance becomes a pattern. A pattern becomes amplification. Amplification becomes compounding interaction. Compounding interaction becomes loss of coordination. And once coordination is sufficiently degraded, system-wide instability is no longer a collection of local problems. It is the emergent behavior of a network that can no longer preserve functional integrity across its own internal connections.

System Interaction

Electrolytes function as an interdependent network. That is the governing mechanism. Sodium, potassium, and magnesium do not operate as separate substances each performing an isolated task within a closed domain. They function through relationship. Their significance lies not only in their individual properties, but in the way those properties are coordinated across compartments, membranes, transport systems, signaling environments, and regulatory responses. Stability does not emerge because each electrolyte is independently "adequate" in some abstract sense. It emerges because their interactions remain coherent enough to preserve gradients, thresholds, fluid distribution, electrical behavior, and system-level coordination at the same time.

This is the necessary starting point because electrolyte physiology is often interpreted through a fragmented model. Sodium is framed as a volume issue, potassium as a muscle or nerve issue, magnesium as a relaxation or support issue, and each is then considered separately as though the body were composed of parallel but largely independent channels. That framing is conceptually convenient, but physiologically misleading. Electrolytes do not merely coexist. They co-regulate. The behavior of one changes the conditions under which the others must function. The meaning of each is therefore relational. To understand any one electrolyte properly, it must be placed within the network of interactions that gives it context.

Sodium and potassium provide the clearest illustration of this interdependence because their relationship across the cell membrane is one of the foundational asymmetries in physiology. Sodium is concentrated primarily outside the cell. Potassium is concentrated primarily inside it. This distribution is not an incidental detail. It is a maintained gradient architecture that supports membrane potential, electrical readiness, signal directionality, and compartmental distinction. The significance of sodium is therefore inseparable from where potassium is, and vice versa. A meaningful description of one immediately implies a relationship to the other. Their functions are not parallel. They are reciprocal.

That reciprocity is what allows gradients to become useful rather than merely descriptive. A sodium gradient only matters because it exists in opposition to other gradients, especially potassium's. Potassium's role in intracellular stability only matters because it exists in structured contrast to extracellular sodium conditions. Electrical signaling, membrane behavior, and transport logic all depend on this asymmetry being maintained not as two independent facts, but as a single coordinated arrangement. If sodium shifts disproportionately relative to potassium, or if potassium becomes unstable relative to sodium, the problem is not simply that one variable has changed. The problem is that the relationship that made both functionally meaningful has degraded.

Magnesium adds a different but equally important layer to this network. It does not mirror sodium or potassium in its role, but it interacts with the conditions they create. Magnesium participates less as a primary determinant of compartmental asymmetry and more as a modulatory influence on how activation, signaling, and control are regulated within that asymmetry. It affects how excitable systems become, how thresholds are managed, how activation is buffered, and how signal behavior remains governable rather than excessive or noisy. This means magnesium cannot be understood as an isolated supportive factor. Its significance emerges through its relationship to systems already shaped by sodium-potassium gradients and the electrical environments those gradients permit.

Taken together, these electrolytes form not a collection, but a coordinated network of dependencies. Sodium contributes to extracellular structure and fluid distribution. Potassium anchors intracellular conditions and membrane behavior. Magnesium influences modulation and control within systems operating on those gradients. None of these roles is self-contained. A disturbance in sodium handling can alter the environment in which potassium-dependent membrane behavior must remain stable. A disturbance in potassium can alter excitability in ways that increase dependence on modulatory control. A disturbance in magnesium can reduce the system's ability to buffer activation generated within already strained sodium-potassium relationships. The network is therefore not additive. It is interactive.

This interactivity creates trade-offs. Trade-offs do not imply that the system is poorly designed or inherently conflicted. They reflect the reality that preserving one domain of stability may require shifting conditions elsewhere. Retaining sodium to preserve extracellular volume may alter broader fluid relationships and influence how concentration is experienced across compartments. Preserving intracellular potassium gradients may require energetic and transport activity that becomes more difficult under broader system strain. Increasing compensatory activation to preserve signaling or pressure may alter the very conditions magnesium-dependent modulation is trying to contain. These are not failures of design. They are the expected consequences of a system whose variables are linked rather than independent.

Trade-offs become especially visible under stress, loss, or constrained regulation. In such states, the system often cannot preserve every variable equally with the same ease. It may prioritize one layer of stability while accepting narrower margin in another. Extracellular conditions may be defended while intracellular strain rises. Signaling may remain intact while recovery becomes less complete. Pressure may be maintained while modulation becomes more burdened. The point is not that the body chooses one electrolyte over another in a simplistic hierarchy. The point is that preserving system coherence under strain often involves shifting dependencies and partial protections rather than perfect maintenance of all domains simultaneously.

This is one reason ratio-based stability matters more than isolated quantity. Ratios express relationships. They capture whether electrolytes remain proportioned in ways that preserve signaling, fluid distribution, and threshold behavior. A system can contain substantial total sodium and still become unstable if sodium's relationship to fluid or to intracellular conditions becomes distorted. Potassium can remain materially present and yet lose functional significance if its relationship to sodium across the membrane begins to narrow or destabilize. Magnesium can remain present while becoming less effective if the ratio between activation pressure and modulatory capacity becomes less favorable. In each case, the instability emerges not from isolated absence, but from relational disruption.

Ratio-based stability should not be reduced to a simplistic numeric formula. It is a broader systems concept. It refers to the maintained proportional relationships through which different electrolytes remain interpretable to the system and useful within it. A stable ratio is one that preserves functional directionality and coordination. An unstable ratio is one that makes signaling less reliable, fluid distribution less coherent, or thresholds less predictable. This is why focusing on one electrolyte at a time can become so misleading. It treats the network as if it were composed of independent components whose behavior can be understood without reference to the others. But physiology does not operate through isolated adequacy. It operates through coordinated relationships.

This is where single-variable thinking begins to fail most visibly. When electrolyte behavior is interpreted one variable at a time, system behavior becomes harder to explain. A person may appear to have "enough" of one electrolyte and still experience instability. Another variable may appear only modestly altered and yet system coordination may deteriorate. These apparent contradictions are often the result of interpreting the network through isolated categories. Once the interdependence is recognized, the contradictions become more intelligible. The issue is not whether one variable crossed a threshold in isolation. The issue is whether the relationships between variables remained coherent enough to preserve the system's operating structure.

This also clarifies why "optimizing" one variable in abstraction is not the same as preserving system stability. A single electrolyte may be elevated, retained, or emphasized in a way that appears favorable when viewed alone, yet still contribute to broader imbalance if it disrupts relationships elsewhere. System stability is not the sum of individually maximized components. It is the outcome of aligned interaction. An independently favorable state for one variable can still be destabilizing if it pushes another relationship outside its useful range or increases compensatory burden elsewhere in the network. What matters is not whether each part appears strong in isolation. What matters is whether the parts still fit together.

This is the deeper meaning of functional coherence. Functional coherence refers to the state in which different electrolyte-driven processes remain aligned enough to support one another rather than interfere with one another. Fluid distribution supports signaling rather than distorting it. Gradients support excitability without producing noise. Modulation contains activation without suppressing necessary responsiveness. Compensation, when present, remains coordinated rather than fragmenting into competing priorities. In a coherent system, the network behaves as a network. Its parts remain distinct, but they still contribute to a shared regulatory logic.

Disruption, by contrast, often begins when this coherence weakens before any single variable appears dramatically abnormal in isolation. Ratios begin to drift. Trade-offs become more costly. Dependencies become more strained. One part of the system starts requiring support from another in a way that narrows overall margin. This may not produce immediate overt instability, but it changes the way the system behaves under load. A network with weakening coherence is less able to absorb stress, less able to preserve signaling consistency, and less able to maintain fluid and electrical conditions across changing demands. The first signs of instability are often relational before they are dramatic.

This relational view also explains why system behavior can feel disproportionate to any one visible change. Because electrolytes function through interdependence, a relatively modest disturbance in one domain can alter the behavior of the whole network if it affects a central relationship. A small shift in sodium handling may become significant if it changes extracellular conditions in ways that alter potassium-dependent excitability and magnesium-dependent modulation at the same time. The disturbance is not "small" in network terms simply because it appears modest in one variable. What matters is whether it changes the architecture of interaction.

Seen clearly, electrolyte stability is not the preservation of three separate substances in parallel. It is the preservation of a network of relationships whose coherence allows signaling, distribution, modulation, and coordination to remain aligned. Sodium, potassium, and magnesium matter not only because of what each does alone, but because of what each allows the others to do within a shared physiological structure. Their trade-offs, dependencies, and ratios are not secondary details. They are the mechanism through which stability emerges at all. When that interdependence is preserved, system behavior remains coordinated. When it begins to fragment, the problem is not that one electrolyte has "gone wrong" in isolation. The problem is that the network itself has become less able to function as a coherent whole.

Context Dependency

Electrolyte dynamics change based on system conditions. That is the governing mechanism. Electrolytes do not behave according to a fixed internal script that unfolds the same way regardless of circumstance. Their distribution, retention, signaling relevance, and regulatory handling are all context-dependent. The meaning of a given electrolyte state changes with fasting, stress, environmental exposure, aging, accumulated load, and adaptive history. Stability therefore cannot be understood through static assumptions alone. What appears stable under one set of conditions may become unstable under another not because the underlying physiology has become inconsistent, but because the conditions governing that physiology have changed.

This is the central problem with static models of electrolyte interpretation. Static models assume that there is a stable and transferable meaning attached to a given quantity, pattern of intake, or apparent level of availability. They imply that if one condition was adequate once, it should remain adequate across contexts. But physiological systems do not operate in abstraction from context. Electrolytes are meaningful only within the regulatory environment in which they are being used, conserved, lost, redistributed, or strained. The same sodium state, the same potassium condition, or the same modulatory burden of magnesium may carry different consequences depending on whether the system is fed or fasted, calm or stressed, cool or heat-loaded, young or operating under narrowed adaptive range.

Fasting is one of the clearest demonstrations of this principle because it changes electrolyte dynamics even in the absence of dramatic external disruption. When intake is removed or compressed, the system does not simply continue operating on the same assumptions with less incoming material. Hormonal conditions shift. Renal handling changes. Fluid distribution may alter. Sodium retention and excretion patterns can change as the body moves through altered metabolic and signaling conditions. Potassium behavior may also be affected indirectly through shifts in membrane dynamics, intracellular-extracellular relationships, and overall regulatory state. Magnesium's modulatory role may become more visible under altered activation or stress states associated with fasting. None of this means fasting is inherently destabilizing in every context. It means it changes the system conditions under which electrolyte behavior must be interpreted.

This is why consistency of intake cannot be treated as proof of consistency of stability. A person may consume or retain similar total quantities across fed and fasted states while experiencing different electrolyte behavior because the context of distribution and regulation has changed. The same material conditions can mean different things when the hormonal and fluid-handling environment is different. This is one of the main ways static thinking fails. It assumes that similar visible inputs imply similar physiological consequences. But electrolyte function depends not only on what is present, but on what state the system is in while that presence is being interpreted and used.

Stress creates another major shift in electrolyte dynamics because it changes both regulatory behavior and system demand. Stress is not relevant here only as a psychological category, but as a physiological condition of altered activation, compensatory burden, and signaling load. Under stress, the system may change how it retains or excretes fluid and electrolytes, how it distributes blood flow and extracellular volume, how it regulates pressure, and how much signaling coherence it must maintain under elevated demand. A system under stress is not simply the same system "with more going on." It is operating under a different regulatory logic, one in which margin may narrow, compensation may increase, and previously tolerable conditions may become less stable.

This matters because stress changes not only how much the system may require from electrolyte-supported processes, but how effectively those processes can still be maintained. Sodium may become more relevant to preserving extracellular conditions under activation-related shifts in volume and pressure behavior. Potassium-dependent excitability may become harder to regulate cleanly if signaling demand rises and recovery becomes less complete. Magnesium's role in modulation may become more burdened as the system attempts to buffer increasing activation and maintain control over excitability and threshold behavior. None of these changes need to be catastrophic to matter. They simply alter the conditions under which electrolyte stability must be preserved.

Environmental variation introduces a different form of context dependency because it changes the external conditions to which internal regulation must adapt. Heat is the most obvious example because it alters sweat-mediated loss, fluid distribution, extracellular demand, and the burden on volume regulation. But environmental context extends beyond temperature. Dryness, altitude, humidity, illness, physical confinement, movement demands, and seasonal shifts can all alter how fluid is partitioned, how much loss occurs, how much compensatory effort is required, and how narrow or wide the system's tolerance remains. Electrolyte behavior is therefore not simply internally determined. It is shaped continuously by the interface between organism and environment.

This environmental dependence explains why the same internal pattern may behave differently in different external conditions. A fluid-electrolyte arrangement that is stable in cool, low-demand circumstances may become much less stable in heat or repeated exposure. A signaling pattern that remains coherent under low environmental stress may become more vulnerable when fluid loss, thermal strain, or altered distribution begins to narrow margin. These are not contradictions. They are the expected behavior of a context-sensitive system. Stability cannot be judged meaningfully without asking what conditions the system is being asked to tolerate while maintaining its gradients, signaling, and compartmental structure.

Age adds another layer because electrolyte dynamics are shaped not only by immediate context but by the system's adaptive range across time. Aging is not best understood here as a moralized decline narrative or as a singular loss of function. It is better understood as a gradual change in regulatory flexibility, recovery speed, distributional resilience, and tolerance for perturbation. A younger system may often preserve fluid and signaling coherence across wider ranges of disruption or recover more quickly from shifts in load. An older system may still remain stable, but often with narrower margin, slower restoration, and less room for repeated or overlapping stressors. The same perturbation can therefore carry different significance depending on the age-related conditions under which it is encountered.

This age-related narrowing does not mean that electrolyte behavior becomes fundamentally different in type. It means it becomes different in tolerance and consequence. A shift in fluid distribution that was once absorbed with little visible disruption may later produce more meaningful instability because the system has less regulatory reserve. A loss pattern that was once rapidly compensated may now take longer to contain. A signaling perturbation that was once buffered easily may now interact more strongly with fatigue, stress, or reduced recovery. Context dependency therefore includes temporal context: not just what conditions exist today, but what kind of system is responding to them.

Adaptation complicates this further in a useful way because systems are not merely passive recipients of context. They respond to repeated conditions over time. Repeated exposure to certain loads can alter handling, tolerance, and pattern recognition within the body. A system repeatedly exposed to heat may alter its fluid and electrolyte handling differently than one without that exposure. A system repeatedly experiencing fasting-related shifts may not respond identically to one encountering them as an acute novelty. A system accustomed to certain signaling or environmental demands may regulate differently than one under unfamiliar load. Adaptation therefore means that context dependency is not only about current conditions, but about the system's history of interaction with those conditions.

This adaptive dimension is important because it prevents context dependency from being misread as instability alone. Variation in electrolyte behavior across conditions is not always evidence of dysfunction. Often it reflects the fact that living systems are responsive and conditional. What matters is not that the system changes with context. What matters is whether those changes preserve or degrade functional integrity under the conditions present. Adaptation can widen tolerance in some domains and narrow it in others. It can increase coherence under familiar loads while reducing flexibility elsewhere. The key point is that electrolyte behavior cannot be understood outside the adaptive history through which current responses are being generated.

This is why no static requirement model can fully account for electrolyte stability. A static requirement model assumes that one stable set of needs, quantities, or rules can explain what is required across all states. But context-dependent physiology resists that simplification. What is relevant under fasting may differ from what is relevant under feeding. What is stable under low stress may not remain stable under chronic activation. What is manageable in one climate or life stage may become less so in another. Even when total intake or gross availability appears unchanged, the system's interpretation and use of those conditions may be entirely different because the surrounding context has changed.

The failure of static models is not merely practical. It is conceptual. Static models tend to treat electrolytes as inventory items rather than as dynamic participants in regulation. They assume that if the material is present, the system should behave predictably. But electrolytes are not just present. They are distributed, retained, lost, shifted, and interpreted through context-sensitive pathways. Their relevance depends on compartmental behavior, hormonal state, environmental load, signaling demand, adaptation history, and recovery capacity. A fixed model cannot fully capture a variable whose meaning changes with the system that is currently using it.

This also clarifies why apparent inconsistency in symptoms or stability is often not inconsistency at all. A person may feel stable under one condition and unstable under another while intake remains relatively unchanged. A previously manageable fasting pattern may feel different under higher stress or greater environmental load. Heat, sleep disruption, aging, repeated activation, or accumulated strain may alter what the same internal arrangement can still support. These changes can appear confusing if stability is expected to follow fixed rules. They become more coherent once context dependency is recognized as the governing principle.

Seen clearly, electrolyte dynamics are not static requirements waiting to be met. They are conditional behaviors unfolding within a living system whose needs, losses, regulatory strategies, and tolerances shift with context. Fasting changes the regulatory landscape. Stress alters retention, demand, and signaling burden. Environment changes loss patterns and fluid behavior. Age and adaptation change margin, response, and recovery. Stability therefore depends less on abstract consistency and more on whether electrolyte behavior remains aligned with the conditions currently shaping the system. When context changes, the meaning of electrolyte stability changes with it, because the system itself is no longer operating under the same terms.

Stability Framing

Electrolytes are foundational infrastructure for system stability. That is the governing mechanism. They are not optional additions layered onto an otherwise self-sustaining organism, nor are they peripheral factors relevant only under special conditions such as heavy sweating, fasting, heat exposure, or visible dysfunction. They belong to the underlying architecture of physiology itself. Electrolytes make gradients possible, and gradients make signaling possible. They shape distribution, and distribution shapes pressure, compartment integrity, and fluid coherence. They support regulation, and regulation is what allows living systems to preserve continuity across changing conditions. To understand electrolytes properly is therefore to understand them not as occasional concerns, but as embedded infrastructure without which physiological coordination cannot be maintained.

This infrastructural framing is necessary because electrolytes are often interpreted too close to the surface. They are treated as accessories to hydration, as nutrients whose importance rises only when symptoms appear, or as context-specific tools used to correct visible disruptions. That framing reverses their actual role. Electrolytes are not primarily responses to instability. They are part of the reason stability is possible in the first place. When they become visible, it is often because the infrastructure they support has become strained enough to enter awareness. But the underlying work was always present. The fact that infrastructure is often unnoticed is not evidence that it is secondary. It is evidence that foundational systems do their most important work in the background.

A useful way to understand this is to recognize that infrastructure is not defined by dramatic output. It is defined by enabling conditions. Roads do not generate commerce directly, but commerce depends on them. Electrical grids do not create thought, but modern systems cannot coordinate without them. In physiology, electrolytes occupy a similar position. They are not the final output of movement, cognition, circulation, or regulation. They are among the enabling conditions that allow those outputs to occur coherently. They are part of what gives direction to fluid, order to signaling, thresholds to activation, and continuity to regulation. Their significance lies less in what they "do" as isolated substances than in what they make possible across systems.

This broader view becomes clearer when the major mechanisms are integrated rather than considered in isolation. Gradients are one such mechanism. Electrolytes create and maintain differences in concentration across membranes and compartments, and those differences generate directionality. Without directionality, there is no meaningful movement, no membrane potential, no controlled signaling, and no coherent partitioning between intracellular and extracellular space. But gradients alone are not the full story. Their functional significance depends on where water goes, how compartments are maintained, and whether the system can preserve asymmetry over time. This is where fluid distribution becomes inseparable from signaling logic.

Fluid distribution is not a separate hydration topic layered onto electrolyte physiology after the fact. It is part of the same infrastructure. Water follows osmotic structure, and osmotic structure is shaped by electrolyte distribution. Extracellular volume, intracellular balance, pressure relationships, and compartment integrity all depend on this interaction. A gradient without distribution is abstract. Distribution without gradients is unstable. The system depends on both at once.

This is one reason fragmented interpretations of electrolytes become so misleading. They separate what physiology keeps integrated. The body does not run sodium as a "fluid issue" and potassium as a "nerve issue" in parallel. It runs a coordinated system in which distribution and signaling are continuously shaping one another.

Electrical signaling emerges from this same integrated infrastructure. Membrane potential is not an isolated electrical fact. It is the electrical expression of maintained electrolyte asymmetry across compartments. Threshold behavior, excitability, signal propagation, and rhythmic coordination all depend on the continuity of that asymmetry. If gradients degrade, signaling degrades. If distribution shifts in ways that distort gradients, signaling degrades. If modulation weakens under load, signaling becomes noisier or less governable. This is why signaling should not be treated as a specialized downstream phenomenon relevant only to nerves or muscles. It is one of the clearest expressions of electrolyte infrastructure at work across the organism.

Regulation then becomes the mechanism that keeps this infrastructure from collapsing under ordinary life. Electrolytes are not simply present in a finished arrangement. They are continuously being retained, excreted, redistributed, rebalanced, buffered, and defended. Stability depends not only on the existence of gradients and compartments, but on the system's capacity to maintain them under changing conditions. That maintenance is regulation. It is the ongoing work required to keep the infrastructure usable. Without it, even a structurally sound arrangement would drift toward disorder. Electrolyte physiology is therefore not merely about composition. It is about active preservation of the conditions that allow composition to remain functionally meaningful.

Seen together, these mechanisms form a single system rather than a list of separate topics. Gradients create direction. Distribution gives that direction physiological structure. Signaling uses that structure to coordinate behavior across tissues and time. Regulation preserves the entire arrangement under changing load. This is what it means to call electrolytes infrastructure. Their role is not confined to one domain because infrastructure by definition underlies multiple domains simultaneously. They support fluid coherence, electrical behavior, compartmental integrity, rhythmic stability, contraction-release cycling, compensatory capacity, and broader system coordination not through separate independent actions, but through one integrated regulatory architecture.

This integrated view also clarifies why stability and optimization are fundamentally different conceptual frameworks. Optimization thinking tends to focus on maximizing, enhancing, or individually improving selected variables. It treats the body as a collection of features to be tuned toward higher performance or more desirable output. Stability thinking is different. Stability asks whether the system remains coherent, coordinated, and functional across time and variation. It is less concerned with peaks than with continuity. In an optimization frame, electrolytes are often treated as inputs to manipulate in pursuit of specific outcomes. In a stability frame, electrolytes are understood as part of the baseline architecture that makes coherent function possible at all.

This distinction matters because optimization logic often distorts how electrolyte physiology is interpreted. If electrolytes are treated primarily as tools for enhancing performance, then their value becomes situational and instrumental. They matter when the user wants more energy, more endurance, fewer symptoms, better output, or sharper response. But this framing makes them seem intermittent and optional. It suggests that electrolyte relevance rises and falls according to the goals or activities of the person involved. A stability frame corrects this by locating electrolyte function where it actually belongs: beneath the level of optional enhancement, inside the level of ongoing system continuity. Electrolytes are not chiefly tools for exceptional moments. They are part of what allows ordinary moments to remain physiologically coherent.

This also explains why electrolyte function should not be understood as intermittent input. Electrolytes are often encountered in discourse at the moment of consumption or replacement, which creates the illusion that their significance begins when they are ingested and ends when they are not being consciously considered. But physiological relevance does not operate on that schedule. Electrolyte-dependent regulation is continuous. Gradients are being preserved continuously. Distribution is being managed continuously. Signaling thresholds are being held continuously. Compensation is occurring continuously when needed. Loss is occurring continuously under varying conditions. The system is not occasionally using electrolytes. It is continuously organized through them.

This continuity changes how instability should be interpreted. If electrolyte function is viewed as intermittent, then disruption appears only when something obvious goes wrong: a cramp, a dizzy spell, a rhythm disturbance, a visible drop in tolerance, a sensation of depletion. But if electrolyte function is understood as infrastructure, then those visible events are not the beginning of the problem. They are moments when the underlying infrastructure has become strained enough to produce readable output. The infrastructure itself was already involved long before the symptom appeared. The visible event is therefore not the entire story. It is a surface-level expression of a deeper continuity that has been under load.

One consequence of this infrastructure model is that it discourages surface-level interpretation. It becomes harder to mistake fatigue for a simple energy issue, dizziness for a generic hydration issue, cramps for a single-substance absence, or rhythm disruption for an isolated event. These may still be valid descriptive labels at the surface, but they no longer suffice as explanations. Infrastructure thinking asks what gradients, what distributional relationships, what signaling conditions, and what regulatory pressures had to change for those outputs to become visible.

It moves interpretation downward, away from isolated events and toward the systems that made them possible.

This deeper framing also makes clear why foundational systems are often overlooked. Surface phenomena are easier to see, easier to narrate, and easier to intervene upon conceptually. Infrastructure is quieter. It operates beneath awareness until its continuity is interrupted. That invisibility creates a recurrent interpretive error: the tendency to focus on visible events while neglecting the structures that made those events either avoidable or inevitable. Electrolytes are especially vulnerable to this error because they are frequently discussed only at moments of overt relevance. But their actual role is broader and more continuous than those moments suggest. They do not become important when instability appears. Instability reveals that their importance was already there.

To frame electrolytes as foundational infrastructure is therefore not merely to elevate their importance rhetorically. It is to place them in the correct physiological category. They belong with the conditions that make regulation, signaling, fluid coherence, and functional integrity possible across time. Their significance is not episodic. It is architectural. Their behavior is not best understood through isolated inputs, isolated symptoms, or isolated interventions, but through the continuity of a system that is always being maintained, always being challenged, and always depending on gradients, distribution, signaling, and regulation to remain coherent.

From that perspective, electrolyte stability is not a niche topic within physiology. It is one of the core ways physiology remains physiologically possible. Stability is not built on top of electrolyte function as a later refinement. It is built through it. The organism remains coordinated not because it occasionally addresses electrolyte problems, but because electrolyte-dependent infrastructure continues operating beneath every signal, every fluid shift, every contraction, every regulatory adjustment, and every attempt to preserve continuity under changing conditions. That continuity is not decorative. It is foundational.

About the Author

Nathan Reed writes about metabolic stability in systems exposed to load, variability, and time.

His work examines how biological capacity is shaped by energy availability, recovery architecture, sleep amplitude, muscle mass, environmental demand and workload. Rather than treating nutrition, fasting, sleep, and training as isolated tactics, he approaches them as interacting regulatory systems governed by margin, buffering, oscillatory coherence, and constraint dominance.

The Stability framework explores how physiological compression accumulates gradually and how proportional calibration preserves long-horizon resilience across decades. The objective is not peak optimization, but preservation of functional continuity under load.

Electrolytes is part of a broader metabolic stability ecosystem developed under Perennial Publishing, each volume addressing a distinct vector of biological durability.

The work is designed for readers who prefer structural clarity over motivational rhetoric and long-horizon coherence over short-term intensity.

The Metabolic Stability Series

Metabolic Stability: *A Systems Guide to Blood Sugar, Fasting, and Long-Term Metabolic Health*

(Cornerstone Volume)

Fasting Fundamentals: *A Systems Guide to Metabolic Flexibility, Fat Loss, and Long-Term Health*

(Foundational Volume)

Intermittent Fasting: *Metabolic Timing, Body Composition, and Sustainable Health*

Prolonged Fasting: *A Systems Guide to Duration, Recovery, and Metabolic Limits*

Electrolytes: *A Systems Guide to Hydration, Electrolyte Balance, and Physiological Stability*

Future Titles Include:

Insulin Stability

Sleep and Recovery (Vitality Series)

Strength and Aging (Vitality Series)

Part of a Broader Stability Framework

The Metabolic Stability Series is part of a larger stability ecosystem published by Perennial Publishing.

Related domains include:

Money Stability — structural durability in financial systems

Career Stability — positioning, load tolerance, and transition resilience

Mental Stability — cognitive steadiness under pressure and uncertainty

Emotional Stability — regulatory steadiness in environments that do not slow down

Foundations (Foundational Living Systems) — low-fragility life architecture

Each domain approaches stability as a structural problem: capacity versus load, buffering versus shock, governance versus drift. Together, they form a unified framework for long-horizon durability across the major systems that shape human trajectory.

About Perennial Publishing

Perennial Publishing produces practical systems guides for building stability in an unstable world.

Our books explore the foundations of long-term resilience across health, finance, career, and everyday decision-making. Each title focuses on clear frameworks, durable principles, and practical strategies that help readers build lasting stability rather than short-term fixes.

The Perennial catalog includes series on metabolic health, financial stability, career systems, and personal consistency. While each book stands on its own, they share a common goal: helping individuals create durable systems that support a stable and sustainable life.

For bulk or institutional orders, contact Perennial Publishing.

Made in United States
North Haven, CT
04 November 2026

21768653R00102